Selected Chapters from the Renin-Angiotensin System

Edited by Aleksandar Kibel

Published in London, United Kingdom

IntechOpen

Supporting open minds since 2005

Selected Chapters from the Renin-Angiotensin System
http://dx.doi.org/10.5772/intechopen.77666
Edited by Aleksandar Kibel

Contributors
Jia Zhuo, Xiao Li, Ana Paula de Oliveira Oliveira Leite, Xu Chen, Chunling Zhao, Jianfeng Zhang, Xiaowen Zheng, Gonzalo Ignacio Gómez, Juan C. Sáez, Victoria Velarde, Katy Satué Ambrojo, Ana Muñoz Juzado, Alan Poisner, Gabriel Salvador, Manuela Ciocoiu, Iris Bararu Bojan, Maria Vladeanu, Codruta Badescu, Aleksandar Kibel, Anto Stazic, Djordje Pojatic, Vedran Dambic, Hayder M. Alkuraishy, Marwa S. Al-Niemi, Ali I. Al-Gareeb, Nasser A. Al-Harchan, Azhar H. Al-Kurashi, Nawar R. Hussain

First published in London, United Kingdom, 2020 by IntechOpen
IntechOpen is the global imprint of INTECHOPEN LIMITED, registered in England and Wales, registration number: 11086078, 7th floor, 10 Lower Thames Street, London, EC3R 6AF, United Kingdom
Printed in Croatia

British Library Cataloguing-in-Publication Data
A catalogue record for this book is available from the British Library

Additional hard and PDF copies can be obtained from orders@intechopen.com

Selected Chapters from the Renin-Angiotensin System
Edited by Aleksandar Kibel
p. cm.
Print ISBN 978-1-78985-421-3
Online ISBN 978-1-78985-422-0
eBook (PDF) ISBN 978-1-78985-554-8

For EU product safety concerns:
IN TECH d.o.o., Prolaz Marije Krucifikse Kozulić 3, 51000 Rijeka, Croatia, info@intechopen.com or visit our website at intechopen.com.

Meet the editor

Aleksandar Kibel is assistant professor at the Department of Physiology and Immunology at the Faculty of Medicine, University of Osijek, as well as an internal medicine specialist and cardiology subspecialist at the Department for Heart and Vascular Diseases, Osijek University Hospital Centre, Croatia. He graduated from the Faculty of Medicine, University of Osijek, with an MD degree in 2009 with a maximum possible grade average, conducting experimental research activities during almost all study years and receiving awards that included the country's top scholarship ("Top Scholarship for Top Students"). His research activities include laboratory practicums and research visits at a number of foreign institutions, including the University Clinic Hamburg-Eppendorf (Germany), the Medical College of Wisconsin (USA), Chulalongkorn University in Bangkok (Thailand), etc. He obtained his PhD degree at the Department of Physiology and Immunology with a focus on vascular physiology and pathophysiology and teaches at medical high school, university, and postgraduate doctoral levels. He was a mentor on several diploma works. Prior experience includes work at the Emergency Medicine Department and an internship at the University Hospital Zagreb, as well as volunteer work at the Red Cross. He has published a number of scientific papers and three book chapters, actively participated in many international scientific conferences and research projects at the preclinical and clinical levels, and contributed to pharmacological clinical trials as a subinvestigator. He is guest editor of several special issues and an editorial board member of several international scientific journals.

Contents

Preface

This book aims to provide a selection of chapters reviewing some of the most relevant and interesting physiological and pathophysiological roles of the renin-angiotensin system (RAS) and discussing novel scientific findings.

The complex and convoluted RAS has been investigated for many years and, through rigorous scientific research, many important and previously unknown components and functions of the RAS have come to light. These discoveries have been crucial in the understanding of this system and provide a basis for effective modulation of the system as part of therapeutic strategies for a number of widespread disorders. New studies are continuing to elucidate the RAS and the mechanisms associated with its functions.

The first chapter, written by Dr. A. Poisner, gives a chronological overview of the many years and decades of important research of the RAS as the scientific community expanded on the importance of this physiological system and established a structured portfolio of knowledge and information about it. The chapter is written with a focus on the personal experiences of Dr Poisner. The next chapter by Dr M. Ciocoiu et al. discusses the genomics, proteomics and therapeutic implications of the RAS. In the following chapter by Dr A. Kibel and coworkers, the significance of the RAS in clinical conditions is presented. The role of RAS in clinically relevant contexts, such as hypertension, heart failure, diabetes mellitus and others can be crucial for both pathogenic reasons and because of therapeutic targeting. The emphasis of the role of the RAS in the treatment of essential hypertension is brought further into focus by Dr G.B. de Oliveira Salvador in his chapter. The intratubular and intracrine RAS in the proximal tubules of the kidney and Its roles in angiotensin II-induced hypertension is discussed by Dr X. C. Li and colleagues. Interesting new scientific results regarding other roles and mechanisms connected to the RAS are susequently discussed in chapters by Dr K. Satue et al. and Dr G. I. Gomez et al. The final chapter of the book, by Al-kuraishy et al., deals with the implications of the new pandemic of COVID-19 as it relates to specific aspects of the RAS and discusses a potential link to Dipeptidyl Peptidase-4.

Taken together, the selected chapters in the book discuss relevant scientific knowledge about the RAS and intend to introduce the reader to cutting-edge research of the RAS with an accentuation on the mechanisms on functional/physiological and on the molecular/cellular levels.

Dr. Aleksandar Kibel
Osijek University Hospital,
Croatia

Chapter 1

From Angiotensin to Renin to Prorenin and from the Adrenal to the Kidney to the Placenta and the Lungs: An Historical Journey

Alan Poisner

Abstract

In 1966 I carried out a study on the role of calcium on angiotensin's stimulant effects on the adrenal medulla. Since then I have been studying the renin-angiotensin system (RAS) for over a half-century in a wide variety of biological preparations, while awareness of its complexity has exploded. My journey has involved studies on genes, proteins, organelles, cells, tissues, glands, organs and whole animals. This chapter reviews what my colleagues and I have learned from these different levels of organization and is not meant to be an update on all features of the RAS. My studies have included experiments on: perfused cat adrenal glands; genetic and second messenger control of catecholamine synthesis and secretion from cultured bovine chromaffin cells and from rats in vivo; renin storage and release in the rat kidney and secretory granules; properties of isolated renin, prorenin and renin-like proteins; hormonal and second messenger control of prorenin secretion from human utero-placental tissues; renin/prorenin in a variety of tumors; and the effect of RAS drugs in a rodent model of pulmonary fat embolism. This most recent study has direct clinical application. I conclude with what I have learned about biomedical research and lessons for the future.

Keywords: angiotensin, renin, prorenin, prorenin receptor, renin-angiotensin system, adrenal medulla, lungs, kidney, placenta, chromaffin granules, calcium, fat embolism

1. Introduction

Awareness of the complexity of the renin-angiotensin system (RAS) has increased exponentially since it was initially considered relevant only to hypertension and has led to an explosion of understanding in biochemistry, molecular biology, cell physiology, anatomy, pharmacology, and pathophysiology. I have been involved in studies at all these levels in a wide variety of experimental models in animals and humans for over 50 years. This chapter is a review of what my colleagues and I have learned over the course of this half-century. This is not meant to be an update on all features of the RAS but rather the advances over the years in my personal research journey. It represents almost half of my total research publications.

This has included studies on the effects of angiotensin on the adrenal medulla in intact cat adrenal glands and cultured bovine chromaffin cells, renin storage and release in the rat kidney and secretory granules, properties of isolated proteins, hormonal and second messenger control of prorenin presence and secretion from human utero-placental tissues and renin/prorenin in a variety of tumors. These studies have implicated the RAS in a rodent model of pulmonary fat embolism syndrome (PFE) and showed that drugs acting at different steps in the RAS provided protection, suggesting that this approach could be useful in treating/preventing this potentially fatal condition. Investigating the RAS in many models in animals and humans should increase our understanding of normal and pathological processes and thus improve therapy/prevention of a variety of diseases. For the sake simplicity the term renin will be understood to mean total renin (renin + prorenin) unless specified otherwise. In extrarenal sites, prorenin may be the only one present and it can have some catalytic activity even without processing to the smaller protein renin, especially when bound to its membrane receptor.

2. Angiotensin and the adrenal medulla

2.1 Mechanism of short-term effects on the intact adrenal gland

In the early 1960s, evidence was presented that angiotensin II (Ang) could evoke the release of aldosterone from the adrenal cortex and catecholamines from the adrenal medulla, but the cellular mechanisms had not been completely identified. Since my colleagues and I had been mining the role of calcium in adrenal medullary secretion in response to acetylcholine [1], we decided to examine peptides in our studies and found that extracellular calcium was required for the stimulant effect of Ang [2]. Interestingly, one of the earliest demonstrations of the direct effect of Ang on adrenomedullary hormone secretion was carried out in 1963 in the laboratory of Wilhelm Feldberg at the National Institute for Medical Research in London [3] at which time I was working in the same lab on a different project during my postdoctoral training. Their study and our later one were carried out on the isolated perfused adrenal glands of cats. This is when I gained experience and insight into the value of using intact tissues in experimental studies without disruption of their cellular connections and revealed the immediate effect of treatment with peptides and amines. We could also stimulate the splanchnic nerve in these preparations to more closely simulate the natural signaling condition. Further studies on the role of calcium in stimulus-secretion coupling revealed its fundamental importance in exocytosis in exocrine secretion and neurosecretion [4–6]. This was summarized in several review papers [7–9].

2.2 Mechanisms of long-term effects on adrenal chromaffin cells

The next time that I had occasion to study the effects of Ang was over 20 years later when I was working on sabbatical in the laboratory of my former student, Dr. J.S. Hong, at the National Institute of Environmental and Health Sciences. His lab was interested in the long-term effects of agents on the adrenal medulla (as a surrogate for postganglionic sympathetic nerves) and the potential feedback on enzymes and peptides. In a series of experiments on isolated chromaffin cells, our group found that a stable Ang peptide (S-Ang) increased the secretion and expression of catecholamines and met-enkephalin as well as the mRNA expression of several catecholamine synthetic genes (tyrosine hydroxylase and phenylethanolamine

N-methyltransferase) [10, 11]. The time course of the response to S-Ang showed both short-term and long-term effects and revealed the increased expression of the oncogene c-fos [12] and its role in nuclear stimulation. These changes were mimicked by in vivo stimulation in rats by insulin [12]. These experiments directly implicated intracellular calcium as a second messenger leading to nuclear mRNA synthesis that required a lag time that followed catecholamine release [11].

2.3 Evidence for endogenous generation of Ang in chromaffin cells

We found evidence that endogenously released Ang from chromaffin cells could initiate the secretion of catecholamines and met-enkephalin from bovine chromaffin cells [13, 14]. It is likely that the enzyme responsible for this is renin since renin has been found in the adrenal medulla and in chromaffin cells [15]. Prorenin was not found in these cells in control animals. These results suggest that there may be some autocrine regulation of adrenomedullary secretion mediated by the RAS.

3. Renin and the kidney

3.1 Properties of renin substrate

In the 1970s, we turned our attention to studies on renin and the analytical method of the day was to measure the generation of angiotensin I using a protein or polypeptide substrate. The literature indicated that renin activity from various sources was not inhibited by the usual SH-targeting agents but was potentiated in some cases by the SH-protecting agent dithiothreitol (DTT). Since we intended to find a useful substrate, we first studied the interaction of DTT with renin and/or several renin substrates (angiotensinogen). The commercially substrate available at the time was hog renin substrate but we also prepared a semi-purified bovine substrate. We found that the potentiating effect of DTT was exerted on the substrate (bovine or porcine) and not on renin [16].

3.2 Subcellular storage of renin

Using our sensitive enzymatic assay and radioimmunoassay (RIA) for angiotensin I, we were able to initiate a number of studies on the storage of renin in the kidney using knowledge gained from our previous studies on secretory granules from the adrenal medulla [17] and the posterior pituitary [18]. The goal was to understand secretory mechanism for renin utilizing the secretory granules from the juxtaglomerular cells of the renal cortex. There had been a few studies on storage of renin but no systematic studies to understand their physical properties and the effect of isolation techniques. In our initial study we took into consideration what we had learned about the influence of temperature and pH on other secretory granules and found that crude rat kidney renin secretory granules were more stable when isolated at room temperature (22–25°) than at 0° and were most stable at pH 6.0 [19]. They were also somewhat stabilized by MgATP unlike adrenal chromaffin granules [20]. Later studies with more purified granules confirmed that the granules were more stable at room temperature but were labile if transferred from hyperosmotic density gradient media back to physiological tonicity [21]. When these purified granules were incubated at 37° instead of room temperature they again showed lability when they were subsequently incubated at 0° [22]. In order to avoid the problems with isolation in hypertonic media, we employed density gradients with lower osmotic properties and at room temperature. These granules were stable as long as there were not transferred to 0° media and kept not far from pH 6.0 [23]. Granules prepared in isotonic density gradient media showed two peaks with short term centrifugation that was resolved with longer term preparation, suggesting that renin granules are of two sizes with the same density [23].

3.3 Renin-like activity in the rat *in vivo*

When we began to study renal granules, we became aware of a study of another angiotensin I-generating enzyme that had a lower pH optimum and preferred the tetradecapeptide substrate rather than the protein substrate. They called it pseudorenin [24]. It was found in rat plasma and a wide variety of tissues and in much higher concentrations in the salivary gland and the spleen than in the kidney. Since we could not find any physiological studies of pseudorenin in intact animals, we examined the changes in plasma pseudorenin and renin in rats after nephrectomy and in response to converting enzyme inhibition and beta-adrenergic receptor blockade. We found that, unlike renin, plasma pseudorenin increased after nephrectomy and treatment with propranolol but did change after angiotensin converting enzyme treatment [25]. Later we examined bovine spleen and provided evidence that pseudorenin is cathepsin D [26].

4. Renin in human tumors

4.1 Enzymatic analysis

Although there had been several reports of increased renin activity in serum or tissues of patients with renal tumors using bioassays for analysis [27, 28], none had followed the clinical course and biochemical evaluation of the patients and utilized agents interfering with angiotensin to modulate the course of the disease. We reported on a patient with bilateral Wilms tumor (nephroblastoma) who exhibited congestive heart failure, hypertension and elevated serum renin using a more contemporary radioimmunoassay and an international standard [29]. We found the patient's clinical course and tumor size in response to surgery and chemotherapy were paralleled by serum renin concentrations (PRC) and his hypertension was ameliorated by saralasin, a peptide angiotensin receptor antagonist (this was before non-peptide ARBs were available). We were able to assay samples with exceedingly high renin concentrations that would not be possible with simple plasma renin activity (PRA) assays since these high dilutions would reduce the available endogenous substrate to suboptimal levels. For our assays, we used substrate from nephrectomized sheep that was known to be a better substrate for human renin. Plasma renin concentration was over 4600 μU/ml before therapy (normal 30–90 in our lab) and fell to 69 after chemotherapy and surgery. A few months later the tumor size increased and so did the renin concentration [29] A partially resected tumor mass was found to contain renin by immunohistochemical and biochemical analysis (1245 μU/g).

4.2 Immunohistochemical analysis

To utilize a non-enzymatic method to localize renin, we utilized a specific antirenin antiserum and examined several non-renal tumors. In preliminary studies we found renin and prorenin in complete and incomplete hydatidiform moles [30, 31]. By 1990, prorenin was a widely accepted name for what had one time been called "inactive renin" and we reported on its presence in cyst fluid and ascites in patients with ovarian tumors [32].

5. Renin/prorenin in human amniotic fluid and amnion

5.1 Purification and properties

When considering the possibility of alternative forms of renin, we noted the report of high concentrations of renin in human amniotic fluid [33] and sought to purify the enzyme from this source. We noted that the original description of renin in amniotic fluid included a step of acidification and that it was subsequently found that this caused an activation of "inactive renin". We compared chromatographic and kinetic properties of endogenous renin, acid- and pepsin-activated renin using bovine and hog substrate and found differences between acid- and pepsin-activated renin [34]. Further purification of the inactive renin allowed separation from the pseudorenin mentioned above that was similarly inhibited by pepstatin [35, 36]. We developed an assay utilizing a single tube for renin-generated angiotensin I and the subsequent radioimmunoassay which greatly facilitated these studies [37]. We also showed that both prorenin and active renin were inactivated by ethyl diazoacetylglycinate, a compound known to inactivate aspartyl proteases but not pepsinogen [38].

5.2 Localization, synthesis and release from amnion

We initially demonstrated the presence of the renin and prorenin with both acid and trypsin activation using nephrectomized sheep plasma substrate. This showed, like amniotic fluid, that the bulk of the potential angiotensin-generating activity was in the inactive (IR)-prorenin form (about 70% in these samples) [39]. Our immunofluorescence study, using antiserum to human kidney renin, showed that the positive cells were the amniotic epithelial cells and not contaminating chorionic cells [39]. We noted at the time that early initial attempts to show synthesis by cultured amniotic cells were negative [40]. Those studies included bioassays of samples that had initially been treated at pH 3.0 so some prorenin would have been activated and recorded as renin [40]. In our study where we did not expose samples to low pH and used trypsin activation to assay IR, we found no IR or R in medium from cultured amniotic cells although similarly cultured chorionic cells produced enormous quantities of IR/prorenin that sometimes required a dilution of 1000× to bring the samples into the assay range [41].

In another model, to assess the potential synthesis of IR from the amnion, we superfused separately the amnion and the chorion from a clamped fetal membrane in a Ussing chamber device. We found that there was a dramatic and increasing release from the chorion side but a low and decreasing amount from the amnion side [41]. Furthermore, when a high concentration of IR previously released on the chorion side, was superfused on the chorion side, there was no increase on the amnion side, thus excluding leakage or transport [41]. Since there are many sources of prorenin in the human uteroplacental complex at term pregnancy [42], there could be more than one source of its presence in amniotic fluid. That could include uptake from fetal urine since prorenin has been found in urine [43] and an uptake system in amnion has been reported [44]. The latter group used amnion explants and found very low levels of renin mRNA and extremely low levels of prorenin protein release but considered that the decidua could be the source of amniotic prorenin. There was no evidence of *de novo* synthesis in our experiments using undigested amnion [41].

6. Prorenin production and release from human chorion and chorionic trophoblasts

6.1 Purification and properties

After determining that the amnion was not a likely source of prorenin in the amniotic fluid, we turned our attention to the other major fetal membrane, the chorion, and more specifically the chorion leave (free chorion). In our initial report [45], we noted the early work that suggested that Hofbauer cells (fetal macrophages) and not trophoblasts or fibroblasts were the renin-containing cells, using the Bowie stain that showed renin in the kidney We used an immunofluorescence technique and employed an antiserum to a highly purified renin preparation from human kidney that we subsequently realized recognized both renin and prorenin. We found that the renin immunoreactivity in the chorion was strictly localized to the cytotrophoblast layer [45]. Subsequent biochemical studies showed that the "inactive renin" from chorion and culture medium from chorionic cells is definitely prorenin [46]. In addition, we found that this same cell layer was positive for hCG [47] and there was a relatively constant ratio of renin/hCG in purified chorionic cells: 5.14 μU renin/mIU hCG. Here and elsewhere the terms renin and prorenin and IR are used interchangeably except where noted since all or almost all the renin is present as prorenin.

6.2 Release of prorenin and hormones from intact membranes and isolated cells

We used our superfused membrane, preparation to show that both prorenin and hCG were released at a constant or increasing rate even after 80 minutes [47]. An interesting finding from our studies on the superfused chorion was the short term release of prorenin by angiotensin II [48]. We also found that angiotensin induced the release of LHRH (GnRH)-like activity from this preparation [49]. In parallel experiments we demonstrated specific binding of angiotensin II to these cells [49].

Since the purified chorionic cells could be grown in tissue cultures for many days, we were able to examine factors that might modify synthesis and release in the short or long term. We could even grow these cells for periods up to 3 months without them losing their capacity to synthesize and release prorenin [41, 48]. When we examined these cells for the steroid hormone progesterone that had been reported to be present in the chorion, we found they indeed did contain progesterone and its synthesis and release could be promoted by various precursor steroids. (pregnenolone and 25HC) [50]. The amount of progesterone released greatly exceed the amount initially found in these cells and the synthesis and secretion were both promoted by agents acting to raise cyclic AMP (cAMP) [50]. These included dibutyryl cAMP, methyl isobutyl-xanthine (MIX), forskolin and cholera toxin. Prorenin secretion by these cultured cells was also promoted by MIX and cholera toxin and especially by cholera toxin in the presence of phorbol myristate acetate (PMA), a protein kinase C agonist. In some cases, the concentration after 72 hours of incubation with these agents reached 700,000 IU/ml. This was not due to an increase in cell numbers since these were confluent monolayers [48]. The dramatic potential for term chorion to synthesis and release prorenin clearly differentiates the secretory process from the renal secretion of renin where it is stored primarily in dense secretory granules and is presumably released by conventional regulated exocytosis. We have found prorenin in the chorion to be mostly in the cytoplasmic fraction of the tissue and not in particulate fractions [49].

7. Prorenin production and release from decidua

7.1 Primary and second messengers

After a report in 1989 that decidual cells have the capacity to synthesize and release active and total renin from decidua [51], we turned our attention to the maternal portion of the feto-placental unit. The first potential hormonal/primary messenger that we examined was relaxin [52]. This hormonal messenger was known to be present in chorion and decidua [53]. We found that renin released from cultured purified decidual cells was 95% prorenin when we did not expose the samples to acidification as was done in the earlier report [51] and that relaxin caused a dose-dependent increase in release that was paralleled by an increase in tissue prorenin and was inhibited by cycloheximide [52]. This was consistent with new protein synthesis. We cited the views at the time on relaxin's potential effects on uterine ripening and decidual prolactin release. This was one of many pieces of our studies on the utero-placental complex that pointed to paracrine or autocrine effects in supporting local autonomy.

The next potential positive regulator of prorenin release that we examined was endothelin since it was known to be present in the placenta and had been found to modify renin release from the kidney. We found that several endothelin peptides caused a dose-dependent release of prorenin that was associated with an increase in renin mRNA [54]. The release was greater than the control content and was not associated with the release of cellular prolactin. This was another clear example of prorenin secretion by the protein synthesis-dependent constitutive secretion. Further studies on the effect of endothelin (ET-1) on prorenin release showed a clear difference from the control of prolactin (PRL) release The calcium ionophore A-23187 stimulated basal prorenin release and potentiated ET-1 stimulated release while having no effect on PRL release; and the calcium channel blocker nifedipine blocked the effect of ET-1 on prorenin but had no effect on PRL [55]. The protein kinase C agonist PMA stimulated basal and potentiated ET-1 induced prorenin release but inhibited basal PRL release and potentiated the inhibitory effect of ET-1 [55]. Finally, the PKC inhibitor staurosporine increased basal PRL release and reversed the inhibitory effect of ET-1 on PRL release. These results indicate that prorenin and PRL release from decidua are affected in different directions by protein kinase C and that prorenin release is dependent on extracellular calcium but PRL release is not [55]. In addition to protein kinase C and calcium, we also studied the influence of cyclic AMP (cAMP). We found that agents which elevated cAMP in decidual cells also stimulate Pro release. These included forskolin, cholera toxin (CT) and dibutyryl-cAMP [56]. Ninety-eight percent of the renin was in the form of Pro. PMA potentiated the effects of CT and dibutyryl cAMP. These studies had therefore implicated cAMP as well as protein kinase C as second messengers in Pro release from decidua.

7.2 Cytokines and prorenin secretion

After a report that lipopolysaccharide (LPS) and tumor necrosis factor-α (TNF) stimulated prostaglandin production by decidua [57], we examined the effects of these agents on prorenin release from our semi-purified decidual cells. We found that LPS inhibited the synthesis and release of both Pro and PRL from the decidual cells in a time and dose-dependent manner [58]. We noted at the time that the inhibitory effect of LPS might be mediated by the release of cytokines from macrophages and then a paracrine effect on stromal cells could ensue.

We also indicated that it could also be due to a direct effect on the stromal cells. We followed up this study by an examination of the effects of two other cytokines, TNF and interleukin-1β (IL-1β). We reported that these cytokines inhibited synthesis and release of renin from cultured decidual cells in a dose-dependent manner [59] and noted that the cells that were initially plated were composed of 22% macrophages (CD-68-staining) and 78% PRL positive (the other major cell type in decidual cells). We therefore concluded that the effects of these two cytokines could have been mediated by their known actions on macrophages. There was no inhibition of DNA synthesis or cell number. It was of interest that the effects of these cytokines was opposite to those on the rat renal tissues where there was an increase in renin release which occurred in minutes [60] unlike the decidual release that took days [59]. The third cytokine that we examined was interferon-ϒ (IFNϒ) which was known to have receptors on placental cells. We found that IFNϒ inhibited Pro release and its mRNA expression in decidual cells. When we employed an additional step of purification using immunomagnetic beads to separate the macrophages, we found that renin release from both populations of cells was inhibited by IFNϒ and TNF and the combination of these two cytokines was even more effective in producing inhibition of release. Since IFNϒ mRNA was found only in the macrophage population, while the IFNϒ receptor was found on both, it suggested that the effect of locally produced IFNϒ on renin release from macrophages could result from both autocrine and paracrine mechanisms while effects on stromal cells would be paracrine in nature [59].

7.3 Release of prorenin from macrophages and monocytes

Since we knew that macrophages represented a significant portion of decidual cells at term pregnancy and represented about 22% of our decidual cell preparation, we decided to examine directly whether these cells could also could be a source of prorenin. We utilized a method employing immunomagnetic beads after coating the macrophages with HLA-DR antibody to separate the macrophages from stromal cells. This increased the portion of HLA-DR (+) cells from 22 to 93%. The purified cells no longer showed mRNA for prolactin which was abundantly expressed in the non-macrophage population [61]. These cells stained for renin with a specific antibody, expressed renin mRNA and released prorenin into culture medium during 3 days of culture. They did not release prolactin. Importantly, the non-macrophage cells also stained positively for renin and released the same amount of renin per ug DNA per cell as the HLA-DR (+) cells. They also did not stain for a cytokine receptor that was present in the macrophage fraction [61]. These results indicated that both types of decidual sells had the capacity to synthesize and release prorenin and strengthened the case for possible autocrine/paracrine signaling. In addition, we collected some peripheral blood monocytes and demonstrated that they also showed mRNA for renin and speculated on some potential functions of the RAS within the uteroplacental complex [61].

7.4 Regulation of renin expression and secretion in differentiated monocytic cells

To study the expression and regulation of renin in a pure cell line, we employed the well-studied U-937 cells which can be differentiated into a terminal macrophage/monocyte phenotype using phorbol dibutyrate (PDBU). We found that the treatment did cause a morphological change that was identical to those reported in the literature [62, 63]. The differentiated cells expressed renin mRNA and released prorenin into culture media [64]. We first looked at the potential regulation by cAMP, which we had found to be important in prorenin release from decidual and placental cells [56, 65] and others had found important in renal juxtaglomerular cells [66]. Renin mRNA and prorenin release were increased by dibutyryl-cAMP, and forskolin. In addition, terbutaline, a β_2-adrenergic agonist known to increase c-AMP, also increased expression and release of prorenin [64]. The stimulation by terbutaline was potentiated by a type IV c-AMP-phosphodiesterase (PDE) inhibitor. It was known that these cells possess β_2-adrenergic receptors and the type IV PDE. The stimulatory effect of terbutaline on renin secretion was inhibited by an angiotensin receptor agonist and also by TNF and LPS+ IFNϒ [64]. Taken together with our studies on isolated decidual macrophages [61], these results reinforced the potential importance of some components of the RAS in the function of macrophages and other bone marrow-derived cells. They also highlight the possible positive and negative autocrine actions of local mediators.

8. Studies on villous placenta

8.1 Gestational differences in the RAS in placentas

The villous placenta at term has very low concentrations of renin with higher concentrations in decidua and chorion [67, 68]. We hypothesized that the renin concentration might be influenced by gestational age as influenced by alterations in hormonal milieu and found that this was indeed the case [69]. We found that there were dramatically high levels of prorenin and active renin in first-trimester pregnancies: prorenin was 1130 μU/mg protein in the first trimester vs. 5.9 at term; the corresponding values for active renin were 330 vs. 0.15. As might be expected, the values for hCG in the first trimester were also greater than at term (2396 vs. 38.6 ng/mg protein). However the levels of hCG and prolactin in decidua did not change much during gestation and there was no detectable prolactin at any stage in placenta [69]. Placental prorenin correlated with chorionic gonadotropin but not prolactin in both groups. and could reflect similar cellular origins.

8.2 Experimental preparations to study the RAS in human placenta *in vitro*

An early preparation that we used was a superfused placental mince that allowed investigations of mostly intact cells with normal cellular contacts over a period of many hours. With this preparation we showed that there was a dramatic increase in prorenin release beginning after 12 hours, reaching levels of 16 μU/ml at 26 hours from a basal level about 0.5 or less [70, 71]. This spontaneous increase was blocked by cycloheximide and actinomycin D, supporting the conclusion that it required new mRNA and protein synthesis, like our results on decidua and chorion. We also showed that the spontaneous release could be amplified by treatment with relaxin [72]. Further evidence of the increase in synthesis of prorenin was found when we measured the tissue content of superfused placental minces after superfusion for 24 hours with or without the adenyl-cyclase stimulator forskolin. This model was useful for rapid kinetic measurements, but the disadvantage was interruption of much cellular connections.

Another model that we used was the dually perfused human cotyledon which allowed nutrients and drugs to reach cells through vascular channels and permitted assessment of vascular reactivity. It was known that AI and AII produced dose-dependent pressor responses which were blocked by the angiotensin antagonist saralasin and the response to AI was blocked by captopril [73]. We showed in this preparation that there was no renin released into the fetal circulation but there was consistent release into the maternal circuit [74]. It was all prorenin.

This preparation suggested that renin in fetal circulation *in vivo* was not coming directly from the maternal vasculature. The advantage of this model was the greater integrity of the *in vitro* system, but it was restricted by logistical considerations to the number of different preparations that could be set up at one time.

The most useful model that we employed retained much cellular connections and could be used over longer periods of time. That was based on an early model of placental explants, sometimes called organ culture [75]. We found the optimal conditions by putting the explants on top of wire-mesh platforms and keeping the fluid level at the surface of the tissue. With this model we examined potential primary and secondary signals in regulation of the placental RAS.

8.3 Prorenin secretion: primary messengers and modulators

It was known in humans that renal renin secretion was stimulated by catecholamines and selectively by β-1 adrenergic agonists [76] and that the villous placenta had both β-1 and β-2 adrenergic receptors [77]. When we studied the effects of epinephrine and beta-adrenergic agonists on placental renin secretion from placental explants, we found that both β-1 and β-2 adrenergic agonists elicited renin secretion, associated with an increase in synthesis [78]. Again, this was about 95% trypsin activatable and presumably prorenin. This is consistent with the view that extrarenal renin in the human reproductive track is almost exclusively prorenin [54, 79]. We discussed the likelihood that beta-agonist-induced renin secretion would be regulated by activators produced by the fetus [78]. At the same time, we found that hCG secretion was selectively stimulated by the β-2 adrenergic agonist terbutaline and that its stimulant action was blocked by a selective antagonist. We showed that the stimulant effects of beta-adrenergic agonists on both renin and hCG secretion were potentiated by selective inhibitors of phosphodiesterase types III and IV [78]. The differences in agonist selectivity between renin and hCG secretion was consistent with findings on their respective localization in term placenta, with renin in cytotrophoblasts and hCG in syncytiotrophoblasts [80].

The likelihood that renin and hCG are released in close proximity to one another suggests that there might be some paracrine regulation involved. With this model system we provided evidence that hCG stimulates renin secretion and tissue levels [81]. Furthermore, the stimulation was potentiated by phosphodiesterase inhibitors, just like renin secretion, and was accompanied by an increase in media cAMP. The effect of hCG was markedly attenuated by the protein kinase A inhibitor H-89. These results suggested that placental renin secretion may be regulated in part by hCG and mediated by cAMP transduction mechanisms [81]. Further support for the influence of cAMP on renin secretion is presented in Section 8.4.

A possible negative regulator on renin release, based on what has been found in studies on cultured juxtaglomerular cells (JG), is angiotensin II, where a purported negative feedback loop reveals an inhibitory action [82]. We found that a stable analog of angiotensin II inhibited the spontaneous release of renin from placental explants during the 72 hour incubation [71, 83]. This paralleled a study where transfected JG cells released prorenin and not renin [82].

Other negative regulators of placental renin secretion that we identified included LPS and the glucocorticoid dexamethasone. They both inhibited spontaneous and stimulated renin release [71]. These agents act at many different sites, including macrophages, so their influence on renin secretion is complicated.

8.4 Prorenin secretion: second messengers and nuclear signals

We had already identified cAMP as a second messenger for chorion and decidual renin release and we examined if similar mechanisms existed in placental prorenin secretion. Evidence for cAMP mediation was supported by our finding that renin release from explants was stimulated by several orders of magnitude by forskolin and by cholera toxin (CTX) [83]. The effects were potentiated by a cAMP phosphodiesterase inhibitor. The enhanced release of renin was accompanied by an increase in hCG found in the media. It is important to note that prolactin was not detected in the media, thus excluding decidual contamination. This was also supported by the fact that the phosphodiesterase inhibitor did not cause an increase in renin release from similarly-treated decidual explants [84]. Not only did the tissue levels of renin increase indicating new synthesis, but there was a decrease in LDH leakage demonstrating cellular integrity [83]. An interesting finding was that an angiotensin II agonist inhibited both the spontaneous and the CTX-enhanced release of renin. This effect was blocked by an angiotensin receptor antagonist. Further evidence on the role of cAMP came from our studies on cAMP-dependent protein kinase (cAPK) [85]. We found that the dobutamine-stimulated secretion of renin and cAMP was accompanied by an increase in tissue cAPK. We used substituted analogues of cAMP, selective for binding sites on cAPK, and found that site B analogues which bound to catalytic or regulatory sites were stimulants of renin secretion but that site A analogs were not [85]. Strengthening the case for the role of cAPK in dobutamine-induced renin secretion, we found that the specific cAPK inhibitor H-89 blocked secretion and an activator SP-cAMPS stimulated secretion [85]. We then used molecular biology techniques to assess the role of mRNA synthesis in the stimulation of renin secretion, using the β-1 agonist dobutamine and the β-2 adrenergic agonist terbutaline. These agents both increased renin mRNA in a dose-dependent manner which paralleled their effects on renin secretion and tissue levels [86]. The effects on renin secretion and tissue levels were blocked by cycloheximide (a translational inhibitor in protein synthesis) and actinomycin D (a transcriptional inhibitor which acts directly on DNA). Actinomycin D blocked the increase in renin mRNA but cycloheximide did not, thus showing the specificity of these agents and the importance of gene regulation in adrenergic stimulation of placental prorenin secretion [86].

Other second messengers that have been studied in a wide variety of tissues including the placenta include eicosanoids, protein kinase C and calcium. Prostaglandins have been known to influence renal secretion of renin and are actively secreted by utero-placental tissues. We reported that meclofenamate, a relatively selective inhibitor of cyclooxygenase, inhibited the release of renin from placental cells and (unpublished studies) that it also inhibited the ET-1 induced renin release from decidua [71]. As with the decidua, the protein kinase C agonist PMA increased renin release from placental explants and the enzyme inhibitor staurosporine was inhibitory [71]. The influence of calcium was contrary to many, but not all, studies on the kidney where calcium is considered an inhibitory messenger on renin secretion. Extracellular calcium caused a dose-dependent increase in short-term renin release between 1.0 and 3.6 mM in contrast to the kidney. It should be noted that there are some studies on renal tissues in special situations where calcium is a positive regulator [87–89] and other studies on extra-renal renin secretion where calcium also has a positive influence on renin secretion [90, 91]. Similarly, angiotensin was shown to have inhibitory effects on the synthesis and release of renin from placental explants but was stimulatory during short term exposure of superfused chorion [48].

These studies on different anatomical portions of the utero-placental complex suggest that local conditions and times of gestation can modulate the regulation of the RAS and that studies on single cells may be missing the complex interactions that exist in vivo.

9. Role of the RAS in pulmonary fat embolism

9.1 Fat embolism model in rats

After 40 years in academia and an enormous expansion of information in the RAS field, I retired. It was short lived because I still wanted to see new developments and help in obtaining information. That is why I was eager to join (and help revitalize) a project on fat embolism that had been dormant for 40 years. Dr. Federico Adler, a retired orthopedic surgeon, asked for my help in restarting a study of fat embolism in rats that he had worked on in the 1960s. My focus was gaining evidence on the potential role of the RAS in fat embolism syndrome, a sometime fatal consequence of long bone fracture (and some other conditions). This brought me back to *in vivo* studies with its advantages and disadvantages. We initially did time- and dose-related studies with intravenous dosing of the triglyceride triolein and used histochemical methods to evaluate pulmonary pathology. We found that there was an early phase beginning very early and peaking at 48 hours [92] with inflammatory, fibrotic and vasoconstrictive effects.

9.2 Effect of RAS drugs on pulmonary fat embolism

Since there were reports that some RAS drugs had beneficial effects in other types of pulmonary injury, we examined the effects of the angiotensin converting enzyme inhibitor (ACEI) captopril and the type 1 antagonist (AT1) losartan when given 1 hour after the triolein. Both agents provided significant protection against the histopathological effects when viewed at 48 hours [93] and provided strong evidence that the acute effects of fat embolism involved the production of angiotensin II and actions on the AT1 receptor. In a later study we found that the pulmonary injury was also ameliorated by the renin inhibitor aliskiren [94].

We subsequently determined that the initial acute phase after fat embolism was followed by a slowly developing smaller inflammatory response and this was associated with an increase in the presence of angiotensin peptides [95]. Since there were still some fat particles present at this later time period, we suggested that one mechanism could be the continued activation of macrophages that were engulfing the fat and signaling mast cells (and perhaps other cells) to release renin and then local angiotensin release. Some support for this view came from two further studies. In one, we gave the AT1 blocker losartan 6 weeks after the triolein injection and examined the rats 4 weeks later. In this experiment the protective effect of losartan was still demonstrated at this late stage, supporting the view of continued activation of the RAS [96]. In another study we found that 24 and 48 hours after triolein there was an increase in renin staining in lungs that diminished but was still present at 3 and 6 weeks [97]. The renin staining increased again at this late stage when the rats were treated with lipopolysaccharide (LPS) [97] which was known to interact with the RAS [98]. Since we had suggested that some of this renin could be in mast cells, we examined the presence of mast cells in triolein-treated rats and the influence of losartan. We found that 10 weeks after triolein there was an increase of mast cells and this was attenuated by losartan [99]. Addition of LPS at 6 weeks caused slightly more mast cells and this was also blocked by losartan. We also considered that macrophages could be a source of renin.

In several of our papers, we have suggested that the RAS drugs could be useful in the treatment or prevention of fat embolism syndrome and there are some other findings of potential clinical interest. The lungs from a pregnant patient who had succumbed to a pulmonary fat embolism were examined at our affiliated hospital and showed the same kind of histopathological changes that we had observed on our rat experiments [100]. Another point of possible clinical interest was our finding that 6 weeks after fat embolism when the animals appeared normal and had grown as well as the saline-treated controls, they were especially sensitive to a "second hit" with LPS [101]. The potential clinical relevance of these findings is that patients who have severe respiratory distress more than would be expected from their presenting diagnosis could be suffering from a "second hit" a long time after a forgotten trauma which has left a smoldering low-grade inflammatory process continuing in the lung. A recent review of our studies on fat embolism syndrome has been published online [102] that implicates the RAS as a key component of this condition.

10. Role of the RAS in homeostasis and lessons for the future

10.1 Historical developments as seen through my research journey

The latest part of my journey related to the RAS (which is still ongoing) comes 53 years after my first paper mentioning angiotensin [2] and 59 years after my first scientific paper as a medical student [103]. Although I have enjoyed working in many areas of biomedical research, it has been very gratifying to see this major part of my career get closer to the long-range goal of improving health care for people in need. An ironic and maybe not surprising development is that my early and long-standing studies on chromaffin granules and their ATPase [17, 20, 104–107] have come full circle with the finding that the granules contain renin and prorenin [15] and their membranes contain the prorenin receptor [108, 109]; and it is now known, but not in the 1960s, that there is a receptor for renin/prorenin (P)RR that can act on second messengers independent of the RAS [110, 111] It was also not appreciated that there was an opposing arm of the RAS that could antagonize many of the deleterious effects of the angiotensin-ACE-AT1 receptor axis. and it is ubiquitous in distribution outside and inside the cell [112–115]. My studies and the current literature suggest that it will be difficult to find any extra-renal system that does not have some components of an endogenous RAS.

10.2 Lessons that I have learned about biomedical research

I will conclude with my views on ways of approaching biomedical research based on my experience and lessons that I have learned.

10.2.1 Pharmacological lessons

Many of these lessons were not appreciated when I was a medical student.

a. Low doses of drugs may have opposite effects of higher doses, the hormetic effect. Sometimes this is because receptors of different sensitivity are activated as the dose is increased. Other times this may be due to non-receptor mediated effects, such as enzyme inhibition. An example in the RAS is the activation of AT2 receptors by angiotensin II opposing the actions on AT1 receptors and the activation of Ang (1–7) receptors as angiotensin II is converted. This is further discussed below under moonlighting (10.2.d.).

b. Species differences: It has long been clear that other species may have differences in metabolism, pharmacokinetics, morphology and a host of features that make direct extrapolation to humans problematic. Also, mice and rats are not equivalent when compared to human biology.

c. Short term vs. long term experiments: Many experiments have shown biphasic response to drugs with opposite effects seen depending on when the observations were made. This is an argument for examining time-response curves in addition to dose-response curves. This has led in the past to some studies missing a response by looking at a single time-point.

d. Response of young animals (organs) may not be the same as that of older ones. This has been obvious for a long time since the changes during maturation in animal biology has long been appreciated. A striking change during development in a single organ has been found in the human placenta as we have noted in our studies on the RAS. Of course, the placenta is a unique organ and has features of many other organ systems, including the liver, kidney and endocrine and nervous systems among others.

10.2.2 Biochemical lessons

Another lesson that was not appreciated is that of moonlighting. This is the area pioneered by the ground-breaking research of Constance Jeffery from 1999 [116] and still expanding in 2018 [117]. It appears that prorenin and the prorenin receptor are archetypes of multi-functional proteins. Prorenin (a) serves as the zymogen precursor of renin that follows cleavage of the prosegment; (b) becomes a catalytically active enzyme without cleavage when bound to its receptor, (c) activates a surface membrane receptor coupled to the generation of intracellular kinases, and (d) likely serves in several capacities intracellularly [118]. This complexity was not imagined at the time we were calling this 'inactive renin'. The prorenin receptor ((P)RR) also has multiple functions, some not related to angiotensin peptides [119].

10.2.3 Experimental lessons from genes to molecules to intact humans

I have carried out experiments using extreme reductionist approaches, such as studies on isolated proteins and on gene expression. Some of our studies were on isolated organelles and on isolated cells. These types of experiments removed many cells from their natural environment and from potential neuronal, paracrine, or endocrine modulation. The next step up in complexity were studies on isolated perfused glands or tissue slices where there was some contact between different cells as in the intact animal but still not complete signaling from the entire organism. Finally, I have studied the effects of drugs on whole animals *in vivo*, including humans. The result of this wide range of studies has provided me with the following perspective. No one approach will give us a complete picture although bringing us closer to a true view of what is going on in health and disease. Our reductionist approach strives to reduce as many variables as possible and each provides some useful information. However, many interactions at higher levels of complexity may be lost or overlooked. Only experiments on normal humans could come close to a real world understanding but that is beyond ethical consideration. In addition to the reservations listed above, our studies do not take into consideration the epigenetic, environmental and social factors that influence how we interact with internal and external stimuli. Some of this newer point of view includes the field of hormesis [120].

Finally, I have been fortunate to have chosen to study the RAS for much of my research career since this system seems to be ubiquitous throughout biology and has only in the past five decades begun to reveal the many ways we depend upon its proper regulation to maintain our health and suffer when it is out of control.

Acknowledgements

I wish to acknowledge the many colleagues who have accompanied (and often educated) me along the way. Their names are cited in the list of references and described in more detail below.

Thanks

Major contributors include the late Edward Walaszek who helped me start my research career as a medical student at the University of Kansas Medical Center (KUMC); the late William Douglas who was my first research mentor and then colleague at the Albert Einstein College of Medicine; the late Stuart Handwerger at Duke University and the Cincinnati Children's Hospital; my students and colleagues Jau-Shyong Hong at KUMC and NIEHS and Gregory Downing at KUMC; and Agostino Molteni at KUMC and Univ. Missouri Kansas City (UMKC) who continues to be my research colleague in studies on fat embolism. My deepest gratitude goes to my late wife Roselle Burstein Poisner who helped me in the laboratory for more than 30 years and provided the ideal partner for family and career.

Author details

Alan Poisner
University of Kansas Medical Center, Kansas City, KS, USA

*Address all correspondence to: apoisner@kumc.edu

References

[1] Douglas WW, Poisner AM. On the mode of action of acetylcholine in evoking adrenal medullary secretion: Increased uptake of calcium during the secretory response. The Journal of Physiology. 1962;**162**:385-392

[2] Poisner AM, Douglas WW. The need for calcium in adrenal medullary secretion evoked by biogenic amines, polypeptides, and muscarinic agents. Proceedings of the Society for Experimental Biology and Medicine. 1966;**123**:62-64

[3] Feldberg W, Lewis GP. The action of peptides on the adrenall medulla. Release of adrenaline by bradykinin and angiotensin. The Journal of Physiology. 1964;**171**:98-108

[4] Douglas WW, Poisner AM. The influence of calcium on the secretory response of the submaxillary gland to acetylcholine or to noradrenaline. Journal of Physiology. 1963;**165**:528-541

[5] Douglas WW, Poisner AM. Stimulus-secretion coupling in a neuro-secretory organ: The role of calcium in the release of vasopressin and oxytocin. Journal of Physiology. 1964;**172**:1-18

[6] Douglas WW, Poisner AM. Calcium movement in the neurohypophsis of the rat and its relation to the release of vasopressin. Journal of Physiology. 1964;**172**:19-30

[7] Poisner AM. Stimulus-secretion coupling in the adrenal medulla and posterior pituitary gland. In: Ganong WF, Martini L, editors. Frontiers in Neuroendocrinology. New York: Oxford Univ. Press; 1972. pp. 33-59

[8] Poisner AM. Mechanisms of exocytosis. In: Usdin E, Snyder SH, editors. Frontiers in Catecholamine Research. New York: Pergamon Press; 1973. pp. 477-482

[9] Poisner AM. Sodium-calcium interaction as a trigger for the secretory process. In: Pharmacology and the Future of Man, Proc. 5th Int. Congr. Pharmacology, San Francisco. Basel: S. Karger; 1972, 1973. pp. 359-368

[10] Stachowiak MK, Poisner AM, Jiang HK, Hudson PM, Hong JS. Regulation of proenkephalin gene expression by angiotensin in bovine adrenal medullary cells. Molecular mechanisms and nature of the second messengers. Molecular and Cellular Neurosciences. 1991;**2**:213-220

[11] Stachowiak MK, Jiang HK, Poisner AM, Tuominen RK, Hong J-S. Short and long term regulation of catecholamine biosynthetic enzymes by angiotensin in cultured adrenal medullary cells. Molecular mechanisms and nature of second messenger systems. The Journal of Biological Chemistry. 1990;**265**:4694-4702

[12] Stachowiak MK, Sar M, Tuominen RK, Jiang HK, An S, Iadarola MJ, et al. Stimulation of adrenal medullary cells in vivo and in vitro induces expression of c-fos proto-oncogene. Oncogene. 1990;**5**:69-73

[13] Poisner AM, Jiang HK, Stachowiak MK, Hudson P, Hong JS. Catecholamine and met-enkephalin secretion from chromaffin cells in response to endogenously released angiotensin. In: 5th Int Symp Chromaffin Cell Biology. 1989

[14] Poisner AM, Jiang HK, Stachowiak MK, Hudson P, Hong JS. Evidence for an endogenous renin-angiotensin system regulating catecholamine and met-enkephalin secretion from bovine adrenal chromaffin cells. Endocrine Society Abstracts. 1989;**71**:468

[15] Berka JL, Kelly DJ, Robinson DB, Alcorn D, Marley PD, Fernley RT, et al.

Adrenaline cells of the rat adrenal cortex and medulla contain renin and prorenin. Molecular and Cellular Endocrinology. 1996;**119**(2):175-184

[16] Poisner AM, Hong JS. Dithiothreitol augments renin activity by an action on renin substrate. Proceedings of the Society for Experimental Biology and Medicine. 1977;**154**:180-183

[17] Poisner AM, Trifaro JM, Douglas WW. The fate of the chromaffin granule during catecholamine release from the adrenal medulla. II. Loss of protein and retention of lipid in subcellular fractions. Biochemical Pharmacology. 1967;**16**:2101-2108

[18] Poisner AM, Douglas WW. Adenosine triphosphate and adenosine triphosphatase in hormone-containing granules of posterior pituitary gland. Science. 1968;**160**:203-204

[19] Hong JS, Poisner AM. Properties of renin granules isolated from rat kidney. Molecular and Cellular Endocrinology. 1976;**5**:331-337

[20] Poisner AM, Trifaro JM. The role of ATP and ATPase in the release of catecholamines from the adrenal medulla. I. ATP-evoked release of catecholamines, ATP, and protein from isolated chromaffin granules. Molecular Pharmacology. 1967;**3**:561-571

[21] Mannisto PT, Poisner AM. Isolation of renin granules from rat kidney cortex by isotonic or hyperosmotic metrizamide-sucrose gradients. Preparative Biochemistry. 1980;**10**:297-316

[22] Mannisto PT, Poisner AM. Further studies on properties of renin granules isolated from rat kidney cortex. Acta Physiologica Scandinavica. 1981;**112**:365-371

[23] Mannisto PT, Poisner AM. Different isotonic density gradients in separation of renin granules from rat kidney cortex. Medical Biology. 1983;**61**:172-178

[24] Skeggs LT, Lentz KE, Kahn JR, Dorer FE, Levine M. Pseudorenin. A new angiotensin-forming enzyme. Circulation Research. 1969;**25**(4):451-462

[25] Eaton DL, Poisner AM. Plasma pseudorenin in rats after alteration in the renin-angiotensin system. Proceedings of the Society for Experimental Biology and Medicine. 1977;**154**:6-8

[26] Johnson RL, Poisner AM. Evidence that pseudorenin activity in bovine spleen is due to cathepsin D. Biochemical Pharmacology. 1977;**26**:2237-2240

[27] Sheth KJ, Tang TT, Blaedel ME, Good TA. Polydipsia, polyuria, and hypertension associated with renin-secreting Wilms tumor. The Journal of Pediatrics. 1978;**92**(6):921-924

[28] Spahr J, Demers LM, Shochat SJ. Renin producing Wilms' tumor. Journal of Pediatric Surgery. 1981;**16**(1):32-34

[29] Stine KC, Goertz KK, Poisner AM, Lowman JT. Congestive heart failure, hypertension, and hyperreninemia in bilateral Wilms' tumor: Successful medical management. Medical and Pediatric Oncology. 1986;**14**:63-66

[30] Poisner AM, Cheng HC, Tomita T, Poisner R, King CR. Report of a hydatidiform mole containing renin, inactive renin, progesterone and HCG. In: 7th Int Congr Endocrinol. 1984. p. 1251

[31] Molteni A, Poisner AM, Lurain J, Poisner R, Brizio-Molteni L. Renin concentrations in complete and incomplete hydatidiform moles. Federation Proceedings. 1987;**46**:843-844

[32] Poisner A, Poisner R, Weed J. Prorenin and renin in cyst fluid and ascites in patients with ovarian tumors. Gynecologic Oncology. 1990;**36**(2):293

[33] Brown JJ, Davies DL, Doak PB, Lever AF, Robertson JIS, Tree M. The presence of renin in human amniotic fluid. Lancet. 1964;**2**:64-66

[34] Johnson RL, Fleming NW, Poisner AM. Chromatographic and kinetic properties of acid- and pepsin- activated inactive renin from human amniotic fluid. Biochemical Pharmacology. 1979;**28**:2597-2600

[35] Johnson RL, Poisner AM, Crist RD. Partial purification and chromatographic properties of inactive renin from human amniotic fluid. Biochemical Pharmacology. 1979;**28**:1791-1799

[36] Johnson RL, Poisner AM. Inhibition of pseudorenin by pepstatin. Biochemical Pharmacology. 1977;**26**:639-641

[37] Poisner AM, Johnson RL, Hanna G, Poisner R. Activation of renin in human amniotic fluid and placental membranes. In: Sambhi MP, editor. Heterogeneity of Renin and Renin Substrate. New York: American Elsevier; 1981. pp. 335-347

[38] Johnson RL, Poisner AM. Inactivation of amniotic prorenin by ethyl diazoacetylglycinate. Biochemical and Biophysical Research Communications. 1980;**95**:1404-1409

[39] Poisner AM, Wood GW, Poisner R, Inagami T. Renin and inactive renin in human amnion at term pregnancy. Proceedings of the Society for Experimental Biology and Medicine. 1982;**169**:4-6

[40] Symonds EM, Stanley MA, Skinner SL. Production of renin by in vitro cultures of human chorion and uterine muscle. Nature. 1968;**217**:1152-1153

[41] Poisner AM, Wood GW, Poisner R. Release of inactive renin from human fetal membranes and isolated trophoblasts. Clinical and Experimental Hypertension. Part A. 1982;**4**:2007-2017

[42] Poisner AM. The human placental renin-angiotensin system. Frontiers in Neuroendocrinology. 1998;**19**:232-252

[43] Pringle KG, Sykes SD, Lumbers ER. Circulating and intrarenal renin-angiotensin systems in healthy men and nonpregnant women. Physiological Reports. 2015;**3**(10):e12586. DOI: 10.14814/phy2.12586

[44] Pringle KG, Wang Y, Lumbers ER. The synthesis, secretion and uptake of prorenin in human amnion. Physiological Reports. 2015;**3**(4):1-7

[45] Poisner AM, Wood GW, Poisner R, Inagami T. Localization of renin in trophoblasts in human chorion laeve at term pregnancy. Endocrinology. 1981;**109**:1150-1155

[46] Higashimori K, Mizuno K, Nakajo S, Boehm FH, Marcotte PA, Egan DA, et al. Pure human inactive renin. Evidence that native inactive renin is prorenin. The Journal of Biological Chemistry. 1989;**J264**:14662-14667

[47] Poisner AM, Cheng HC, Wood GW, Poisner R. Storage and release of renin and HCG in trophoblasts from human chorion laeve. Trophoblast Research. 1983;**1**:279-298

[48] Poisner AM, Poisner R. The use of human chorionic membranes and isolated trophoblasts for studying renin secretion. In: Poisner AM, Trifaro JM, editors. In Vitro Methods for Studying Secretion. Amsterdam: Elsevier; 1987. pp. 155-169

[49] Poisner AM. Storage, processing and release of chorionic renin and LHRH activity. In: Mochizuki M, Hussa R, editors. Placental Protein Hormones. Amsterdam: Elsevier; 1988. pp. 161-170

[50] Tonkowicz PA, Poisner AM. Evidence for a role for adenosine 3′,5′-monophosphate in progesterone secretion by human chorion. Endocrinology. 1985;**116**:646-650

[51] Shaw KJ, Do YS, Kjos S, Anderson PW, Shinagawa T, Dubeau L, et al. Human decidua is a major source of renin. The Journal of Clinical Investigation. 1989;**83**:2085-2092

[52] Poisner AM, Thrailkill K, Poisner R, Handwerger S. Relaxin stimulates the synthesis and release of prorenin from human decidual cells: Evidence for autocrine/paracrine regulation. The Journal of Clinical Endocrinology and Metabolism. 1990;**70**(6):1765-1767

[53] Bryant-Greenwood GD, Rees MC, Turnbull AC. Immunohistochemical localization of relaxin, prolactin and prostaglandin synthase in human amnion, chorion and decidua. The Journal of Endocrinology. 1987;**114**:491-496

[54] Chao H-S, Poisner AM, Poisner R, Handwerger S. Endothelins stimulate the synthesis and release of prorenin from human decidual cells. The Journal of Clinical Endocrinology and Metabolism. 1993;**76**:615-619

[55] Chao H-S, Poisner AM, Poisner R, Handwerger S. Endothelin-1 modulates renin and prolactin release from human decidua by different mechanisms. The American Journal of Physiology. 1994;**30**:E842-E846

[56] Poisner AM, Thrailkill K, Poisner R, Handwerger S. Cyclic AMP as a 2nd messenger for prorenin release from human decidual cells. Placenta. 1991;**12**:263-267

[57] Romero R, Mazor M, Wu YK, Avila C, Oyarzun E, Mitchell MD. Bacterial endotoxin and tumor necrosis factor stimulate prostaglandin production by human decidua. Prostaglandins, Leukotrienes & Essential Fatty Acids. 1989;**37**:183-186

[58] Chao H-S, Poisner AM, Poisner R, Handwerger S. Lipopolysaccharides inhibit prolactin and renin release from human decidual cells. Biology of Reproduction. 1994;**50**:210-214

[59] Jikihara H, Poisner AM, Handwerger S. Tumor necrosis factor-alpha and interleukin-1 beta inhibit the synthesis and release of renin from human decidual cells. The Journal of Clinical Endocrinology and Metabolism. 1995;**80**:195-199

[60] Antonipillai I, Wang Y, Horton R. Tumor necrosis factor and interleukin-1 may regulate renin secretion. Endocrinology. 1990;**126**:273-278

[61] Jikihara H, Poisner AM, Hirsch R, Handwerger S. Human uterine decidual macrophages express renin. The Journal of Clinical Endocrinology and Metabolism. 1995;**80**:1273-1277

[62] Ways DK, Qin W, Garris TO, Chen J, Hao E, Cooper DR, et al. Effects of chronic phorbol ester treatment on protein kinase c activity, content, and gene expression in the human monoblastoid u937 cell. Cell Growth & Differentiation. 1994;**5**:161-169

[63] Hoff T, Spencker T, Emmendoerffer A, Goppelt Struebe M. Effects of glucocorticoids on the TPA-induced monocytic differentiation. Journal of Leukocyte Biology. 1992;**52**:173-182

[64] Jikihara H, Handwerger S, Poisner AM. Beta-adrenergic regulation of renin expression in differentiated U-937 monocytic cells. Biochemical Pharmacology. 1997;**53**:1883-1888

[65] Poisner AM, Poisner R, Joachims B. Cyclic AMP-induced and spontaneous increase in

prorenin synthesis and release from human placenta in vitro. Placenta. 1991;**12**:428

[66] Skott O, Salomonsson M, Persson AEG, Jensen BL. Mechanisms of renin release from juxtaglomerular cells. Kidney International. 1991;**39** (Suppl. 32):S16-S19

[67] Skinner SL, Lumbers ER, Symonds EM. Renin concentration in human fetal and maternal tissues. American Journal of Obstetrics and Gynecology. 1968;**101**:529-533

[68] Lenz T, Sealey JE, August P, James GD, Laragh JH. Tissue levels of active and total renin, angiotensinogen, human chorionic gonadotropin, estradiol, and progesterone in human placentas from different methods of delivery. The Journal of Clinical Endocrinology and Metabolism. 1989;**69**:31-37

[69] Downing GJ, Poisner AM, Barnea ER. First-trimester villous placenta has high prorenin and active renin concentrations. American Journal of Obstetrics and Gynecology. 1995;**172**:864-867

[70] Poisner AM, Poisner R, Joachims B. Synthesis and release of renin and prorenin from superfused human placenta: Spontaneous increase and response to elevated cyclic AMP. The FASEB Journal. 1991;**5**:A1238

[71] Poisner AM. Regulation of utero-placental prorenin. In: Mukhopadhyay A, Raizada M, editors. Tissue Prorenin-Renin-Angiotensin Systems: Local Regulatory Actions in Reproductive and Endocrine Organs. New York: Plenum Press; 1995. pp. 411-426

[72] Poisner AM, Poisner R. Relaxin induces the synthesis and release of prorenin from human placental tissues. The FASEB Journal. 1990;**4**:A962

[73] Hosokawa T, Howard RB, Maguire MH. Conversion of angiotensin I to angiotensin II in the human foetoplacental vascular bed. British Journal of Pharmacology. 1985;**84**:237-241

[74] Maguire MH, Howard RB, Hosokawa T, Poisner AM. Effects of some autocoids on human fetoplacental vascular resistance: Candidates for local regulation of fetoplacental blood flow. Trophoblast Research. 1988;**3**:203-214

[75] Miller RK, Genbacev O, Turner MA, Aplin JD, Caniggia I, Huppertz B. Human placental explants in culture: Approaches and assessments. Placenta. 2005;**26**(6):439-448

[76] Weber F, Brodde OE, Anlauf M, Bock KD. Subclassification of human beta-adrenergic receptors mediating renin release. Clinical and Experimental Hypertension. Part A. 1983;**5**(2):225-238

[77] Bahouth SW, Malbon CC. Human beta-adrenergic receptors. Simultaneous purification of beta 1- and beta 2-adrenergic-receptor peptides. The Biochemical Journal. 1987;**248**:557-566

[78] Downing GJ, Poisner R, Poisner AM. Beta-adrenoceptor activation stimulates, and phosphodiesterase inhibition potentiates, placental prorenin synthesis and release. The Journal of Clinical Endocrinology and Metabolism. 1994;**78**:41-47

[79] Hsueh WA. Renin in the female reproductive system. Cardiovascular Drugs and Therapy. 1988;**2**:473-477

[80] Unnikumar KR, Wegmann R, Panigel M. Immunohistochemical profile of the human placenta. Studies on localization of prolactin, human chorionic gonadotropin, human placental lactogen, renin and oxytocin. Cellular and Molecular Biology. 1988;**34**:697-710

[81] Downing GJ, Maulik D, Poisner AM. Human chorionic gonadotropin stimulates placental prorenin secretion: Evidence for autocrine/paracrine regulation. The Journal of Clinical Endocrinology and Metabolism. 1996;**81**:1027-1030

[82] Pinet F, Mizrahi J, Laboulandine I, Menard J, Corvol P. Regulation of prorenin secretion in cultured human transfected juxtaglomerular cells. The Journal of Clinical Investigation. 1987;**80**(3):724-731

[83] Poisner AM, Downing GJ, Poisner R. Prorenin secretion from villous placenta: Regulation by cyclic AMP and angiotensin. Placenta. 1994;**15**:487-499

[84] Poisner AM, Downing GJ, Poisner R. Phosphodiesterase inhibitors have differential effects on placental versus decidual prorenin release. Proceedings of the Western Pharmacology Society. 1994;**37**:87-88

[85] Downing GJ, Poisner AM. cAPK mediates placental renin secretion stimulated by beta-adrenoceptor activation. The American Journal of Physiology. 1994;**267**:E954-E960

[86] Downing GJ, Yan BF, Poisner AM. Beta-adrenoceptor activation-induced placental prorenin secretion is mediated by increased renin messenger RNA and protein synthesis. Molecular Pharmacology. 1997;**51**:201-208

[87] Chen D-S, Poisner AM. Direct stimulation of renin release by calcium. Proceedings of the Society for Experimental Biology and Medicine. 1976;**152**:565-567

[88] Morimoto S, Yamamoto K, Horiuchi K, Tanaka H, Ueda J. A release of renin from dog kidney cortex slices. Japanese Journal of Pharmacology. 1970;**20**:536-545

[89] Michelakis AM. The effect of sodium and calcium on renin release in vitro. Proceedings of the Society for Experimental Biology and Medicine. 1971;**137**:833-836

[90] Peters J, Munter K, Bader M, Hackenthal E, Mullins JJ, Ganten D. Increased adrenal renin in transgenic hypertensive rats, tgr(mren2)27, and its regulation by camp, angiotensin ii, and calcium. The Journal of Clinical Investigation. 1991;**91**:742-747

[91] Mizuno K, Hoffman LH, McKenzie JC, Inagami T. Presence of renin secretory granules in rat adrenal gland and stimulation of renin secretion by angiotensin II but not by adrenocorticotropin. The Journal of Clinical Investigation. 1988;**82**(3):1007-1016

[92] McIff TE, Poisner AM, Herndon B, Lankachandra K, Schutt S, Haileselassie B, et al. Fat embolism: Evolution of histopathological changes in the rat lung. Journal of Orthopaedic Research. 2010;**28**(2):191-197

[93] McIff TE, Poisner AM, Herndon B, Lankachandra K, Molteni A, Adler F. Mitigating effects of captopril and losartan on lung histopathology in a rat model of fat embolism. The Journal of Trauma. 2011;**70**:1186-1191

[94] Fletcher A, Molteni A, Ponnapureddy R, Patel C, Pluym M, Poisner AM. The renin inhibitor aliskiren protects rat lungs from the histopathological effects of fat embolism. Journal of Trauma and Acute Care Surgery. 2017;**82**:338-344

[95] Poisner AM, Adler F, Uhal B, McIff TE, Schroeppel JP, Mehrer A, et al. Persistent and progressive pulmonary fibrotic changes in a model of fat embolism. Journal of Trauma and Acute Care Surgery. 2012;**72**(4):992-998

[96] Poisner A, Herndon B, Bass D, Fletcher A, Jain A, England JP, et al. Evidence for angiotensin mediation of the late histopathological effects of pulmonary fat embolism: Protection by losartan in a rat model. Experimental Lung Research. 2018;**44**:361-367

[97] Poisner AM, Herndon B, Al-Hariri A, Qin C, Quinn T, Molteni A. Renin as a mediator of pulmonary damage caused by fat embolism and LPS. The FASEB Journal. 2013;**27**:lb444

[98] Li Y, Zeng Z, Li Y, Huang W, Zhou M, Zhang X, et al. Angiotensin-converting enzyme inhibition attenuates lipopolysaccharide-induced lung injury by regulating the balance between angiotensin-converting enzyme and angiotensin-converting enzyme 2 and inhibiting mitogen-activated protein kinase activation. Shock. 2015;**43**(4):395-404

[99] Poisner AM, Hamidpour S, Ho A, Skaria P, Fletcher A, Simon S, et al. Losartan blocks the recruitment of mast cells in the lungs of rats subjected to fat embolism with or without a second hit with LPS. The FASEB Journal. 2016;**30**:700.2

[100] Ajemba O, Zia H, Lankachandra K, Singh G, Poisner AM, Herndon B, et al. Fat embolism syndrome following caesarean section in an obese patient and its histopathological similarity to an animal model of FE: A case report. Case Reports in Clinical Pathology. 2015;**2**:30-35

[101] Poisner AM, Herndon B, Lankachandra K, Likhitsup A, Al Hariri A, Kesh S, et al. Fat embolism sensitizes rats to a "second hit" with lipopolysaccharide: An animal model of pulmonary fibrosis. Journal of Trauma and Acute Care Surgery. 2015;**78**:552-557

[102] Alan M. Poisner and Agostino Molteni. Fat Embolism: What We Have Learned from Animal Models [Online First]. IntechOpen. 22nd March 2019. DOI: 10.5772/intechopen.85178. Available from: https://www.intechopen.com/online-first/fat-embolism-what-we-have-learned-from-animal-models

[103] Poisner AM. Serum phenylalanine in schizophrenia: Biochemical genetic aspects. The Journal of Nervous and Mental Disease. 1960;**131**:74-76

[104] Douglas WW, Poisner AM. Evidence that the secreting adrenal chromaffin cell releases catecholamines directly from ATP-rich granules. Journal of Physiology. 1966;**183**:236-248

[105] Malamed S, Poisner AM, Trifaro JM, Douglas WW. The fate of the chromaffin granule during catecholamine release from the adrenal medulla. 3. Recovery of a purified fraction of electron-translucent structures. Biochemical Pharmacology. 1968;**17**(2):241-246

[106] Poisner AM, Trifaro JM. The role of adenosine triphosphate and adenosine triphosphatase in the release of catecholamines from the adrenal medulla. 3. Similarities between the effects of adenosine triphosphate on chromaffin granules and on mitochondria. Molecular Pharmacology. 1969;**5**:294-299

[107] Trifaro JM, Poisner AM. The role of ATP and ATPase in the release of catecholamines from the adrenal medulla. II. ATP-evoked fall in optical density of isolated chromaffin granules. Molecular Pharmacology. 1967;**3**:572-580

[108] Nostramo R, Serova L, Laukova M, Tillinger A, Peddu C, Sabban EL. Regulation of nonclassical renin-angiotensin system receptor gene expression in the adrenal medulla by acute and repeated immobilization stress. American Journal of Physiology.

Regulatory, Integrative and Comparative Physiology. 2015;**308**(6):R517-R529

[109] Ludwig J, Kerscher S, Brandt U, Pfeiffer K, Getlawi F, Apps DK, et al. Identification and characterization of a novel 9.2-kDa membrane sector-associated protein of vacuolar proton-ATPase from chromaffin granules. The Journal of Biological Chemistry. 1998;**273**(18):10939-10947

[110] Ichihara A. (Pro)renin receptor and vacuolar H(+)-ATPase. The Keio Journal of Medicine. 2012;**61**(3):73-78

[111] Peters J. The (pro)renin receptor and its interaction partners. Pflügers Archiv. 2017;**469**(10):1245-1256

[112] Santos RAS, Oudit GY, Verano-Braga T, Canta G, Steckelings UM, Bader M. The renin-angiotensin system: Going beyond the classical paradigms. American Journal of Physiology. Heart and Circulatory Physiology. 2019;**316**(5):H958-H970

[113] Zhuo JL, Li XC. New insights and perspectives on intrarenal renin-angiotensin system: Focus on intracrine/intracellular angiotensin II. Peptides. 2011;**32**(7):1551-1565

[114] Kumar R, Thomas CM, Yong QC, Chen W, Baker KM. The intracrine renin-angiotensin system. Clinical Science. 2012;**123**(5):273-284

[115] Ferrario CM, Ahmad S, Varagic J, Cheng CP, Groban L, Wang H, et al. Intracrine angiotensin II functions originate from noncanonical pathways in the human heart. American Journal of Physiology. Heart and Circulatory Physiology. 2016;**311**(2):H404-H414

[116] Jeffery CJ. Moonlighting proteins. Trends in Biochemical Sciences. 1999;**24**(1):8-11

[117] Jeffery CJ. Protein moonlighting: What is it, and why is it important? Philosophical Transactions of the Royal Society of London. Series B, Biological Sciences. 2018;**373**(1738)

[118] Batenburg WW, Danser AH. (Pro)renin and its receptors: Pathophysiological implications. Clinical Science. 2012;**123**(3):121-133

[119] Nguyen G. Renin, (pro)renin and receptor: An update. Clinical Science. 2011;**120**(5):169-178

[120] Calabrese EJ. Hormesis: Path and progression to significance. International Journal of Molecular Sciences. 2018;**19**:2871. DOI: 10.3390

Chapter 2

The Renin-Angiotensin-Aldosterone System: Genomics, Proteomics and Therapeutic Implications

Manuela Ciocoiu, Iris Bararu-Bojan,
Maria Vladeanu and Codruta Badescu

Abstract

Since its discovery in 1898, the renin-angiotensin-aldosterone system (RAAS) has been intensely studied in the medical community, which led to important breakthroughs concerning the treatment of heart diseases. The main role of RAAS is to maintain the circulatory homeostasis, by maintaining the fluid volume. Angiotensin II (ANG II) can act on two receptors: angiotensin type 1 and angiotensin type 2 (AT1R and AT2R). The effect of AT1R consists in increased sodium retention, promotes vasoconstriction (mostly on the efferent arteriole), induces sympathetic nervous system activity, determines thirst and promotes the release of aldosterone. Abnormal activation of RAAS will determine hypertension and cardiac hypertrophy that may lead to heart failure. This is the reason why the pharmacological inhibition of this system has proven to induce such a beneficial effect in cardiovascular diseases such as hypertension and congestive heart failure. Later studies of patients with coronary artery disease revealed that angiotensin-converting enzyme (ACE) gene is also involved in the process of atherosclerosis and those mutations in its gene account for an increased susceptibility to severe acute coronary events. The most common ACE gene mutation is represented by deletions and insertions in the 16th intron (presence or absence of the 287-bp Alu repeat sequence), resulting in three possible genotypes, identified by the length of the fragments: II (490 bp), ID (490, 190 bp) and DD (190 bp). Scientific evidence suggests that the D allele plays a major role in the determination of coronary artery disease. The next step would be to develop new treatment strategies according to the genetic background of each patient.

Keywords: renin, angiotensin, aldosterone, sodium retention, hypertension, ACE gene mutations, D allele

1. Introduction

The renin-angiotensin-aldosterone system (RAAS) has long been discovered, since 1898 by Tigerstedt and Berman, but still possesses many mysteries to be solved. The RAAS has a major impact in controlling blood pressure and fluid balance.

The most important result of the activation of this system is the generation of angiotensin II (ANG II), which is a biologically active hormone that is produced through sequential cleavage of peptides derived from the initial substrate: the molecule of angiotensinogen. ANG II induces its actions after binding to different types of receptors and will induce a wide spectrum of biological reactions that can impact almost every system: the immune system, vessels, brain, heart and kidney.

The main role of RAAS is to maintain the circulatory homeostasis, by maintaining the fluid volume. Abnormal activation of RAAS will determine hypertension and cardiac hypertrophy that may lead to heart failure. This is the reason why the pharmacological inhibition of this system has proven to induce such a beneficial effect in cardiovascular diseases such as hypertension and congestive heart failure.

2. Renin-angiotensin-aldosterone system physiopathology

Renin is produced by the epithelioid cells of the juxtaglomerular cells under the form of a precursor named preprorenin. Afterwards it may be released as prorenin or may be processed to active renin, which will be stored in granules. The release of renin granules is the rate-limiting step of the renin-angiotensinogen-aldosterone cascade.

The next step consists in the release of angiotensinogen from the liver, which will be metabolized by renin, thus liberating angiotensin I (ANG I) [1, 2].

Angiotensinogen is included in the superfamily of Serpin A8 proteins. Serpin is a large and diverse superfamily of protease inhibitors and related proteins. It is released into the blood stream after removal of the 33-amino acid signal peptide and will remain in the circulation for approximately 5 h. However, it remains still unclear how much intact angiotensinogen versus catabolized angiotensinogen [the so-called des-(Ang I)-Agt] is present in the circulation. There is few data in the literature describing the proportions of intact Agt versus des-(Ang I)-Agt in the circulation. Other studies have suggested that des-(Ang I)-Agt may induce angiogenesis. Even though the liver is the main source of angiotensinogen synthesis, there are some other sources for this enzyme: the brain, heart, kidney, lung, adrenal gland, adipose tissue, blood vessels and digestive tube. An independent tissue regulation of angiotensinogen levels has also been proven [3–6].

ANG I will be transformed in ANG II due to angiotensin-converting enzyme (ACE), which is released from the endothelial cells. The angiotensinogen-converting enzyme is a dicarboxypeptidase that cleaves two amino acids from ANG I, thus generating ANG II (**Figure 1**).

There have been described two different types of ACE: somatic and testicular, which are both a result of the alternative splicing of a single gene. The role of ACE in forming ANG II from ANG I is primordial for its biological functions as ANG II is the major effector molecule of the RAAS. The ACE acts on other biologically active peptides, not only on ANG I, one of the most representative being bradykinin [7–9].

Bradykinin will be transformed into an inactive peptide through the action of ACE. This biological pathway for bradykinin metabolism is very significant in vivo. The ancient term for ACE is kininase II. As bradykinin promotes vasodilatation and induces a natriuretic effect, the pharmacological inhibition of ACE will diminish the kininase activity and will subsequently lower the blood pressure. The cleavage of ANG II by angiotensin-converting enzyme type 2 produces the heptapeptide angiotensin 1-7 (Ang 1-7). This peptide binds to the Mas receptor (MasR) and induces downstream vasodilatation, which has an opposing effect of the hypertensive action of AT1R signalling. The cardioprotective properties of Ang 1-7 can diminish or reverse heart failure and hypertensive cardiac remodelling [10] (**Figure 2**).

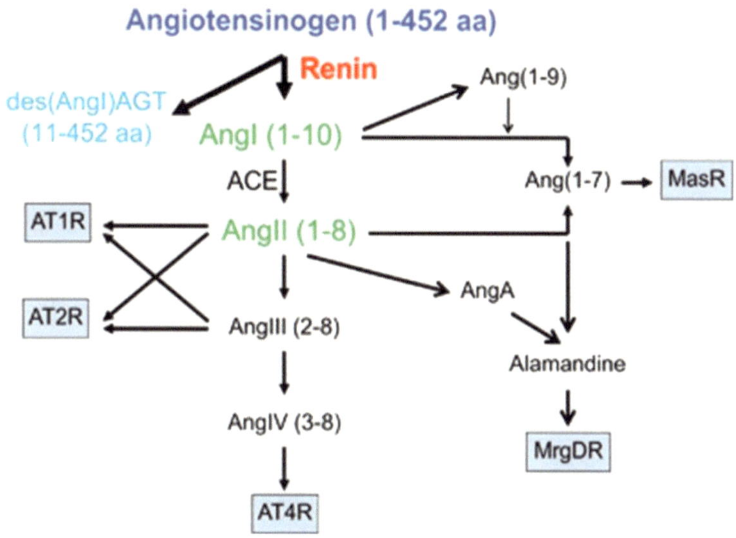

Figure 1.
The renin-angiotensin-aldosterone cascade [3].

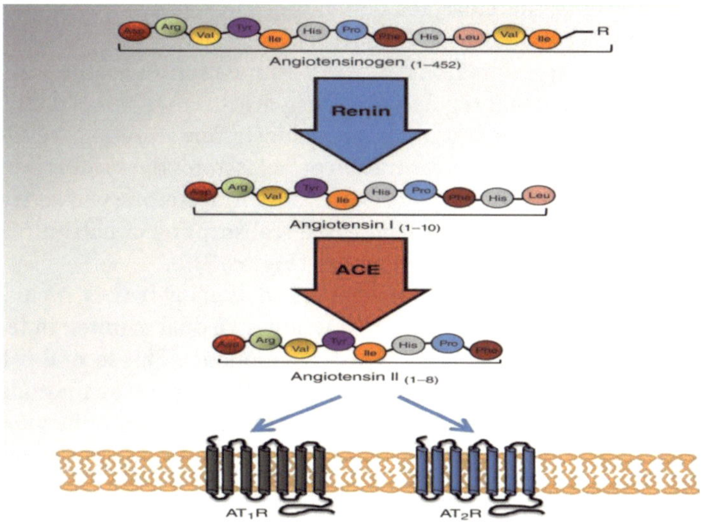

Figure 2.
ANG II formation [11].

ACE2 is a monocarboxypeptidase that transforms ANG I into Ang 1-9 which is a nonapeptide, and ANG II in Ang 1-7 (a heptapeptide). The discovery of these molecules unravels a distinct enzymatic pathway for ANG I and ANG II catabolism that will have an antagonist role to RAAS activation.

Ang 1-7 is a biologically active peptide that induces a wide range of effects, many of them being antagonist to those caused to Ang II. An endogenous orphan receptor, Mas (MasR), was identified in 2003; afterwards this receptor was proved to be Ang 1-7 receptor. It also protects cardiovascular function as it enhances vasodilatation via elevated release of NO and bradykinin, as well as diminishing the production of reactive oxygen species (ROS). The major effects of Ang 1-7 are those induced by MasR activation. These effects counterbalance the ones induced by ANG II and include the activation of the phosphatidylinositol 3-kinase

(PI3K)-Akt-endothelial nitric oxide synthase (eNOS) pathway, the inhibition of protein kinase C (PKC)-p38 MAPK pathways and the inhibition of collagen expression to limit cardiac fibrosis [12].

Also Ang 1-7 therapies have shown cardioprotective effects in preclinical models of non-ischemic and ischemic cardiomyopathy as it inhibited cardiomyocyte growth in vitro and diminished the ventricular hypertrophy induced by myocardial infarction in vivo. Ang 1-7 diminished the myocardial levels of pro-inflammatory cytokines (TNFα and IL-6), thus having a beneficial effect in cardiac inflammation.

Recent studies demonstrated that recombinant human ACE2 generated Ang 1-7 and Ang 1-9, while recombinant murine ACE2 generated predominantly Ang 1-7.

Ang 1-9 has also proved to be beneficial as it acts on AT_2R, thus leading to cardio protection. Therefore, the ACE2/Ang 1-7/MasR and ACE2/Ang 1-9/AT_2R axes are now considered to be physiological antagonists that inhibit the RAAS [10, 12, 13].

ANG II can act on two receptors: angiotensin type 1 and angiotensin type 2 (AT1R and AT2R). The effect of AT1R consists in increased sodium retention, promotes vasoconstriction (mostly on the efferent arteriole), induces sympathetic nervous system activity, determines thirst and promotes the release of aldosterone from the glomerular zone of the suprarenal glands (**Figure 3**).

The action on AT2R stimulation are antagonist to the one on AT1R as it induces vasodilatation and inhibits the inflammatory and fibrotic processes. This receptor is prevalent in the foetus; therefore, it has an important role in normal ontogenesis but has a minor role in adults [13–17] (**Figure 4**).

Aldosterone secretion results from the AT1R stimulation and is the final step of the RAAS. It is an important regulator of the hydric balance and of sodium and potassium exchange. The strongest inducers in aldosterone secretion are ANG II and the increased extracellular concentration of potassium, the synthesis of aldosterone being induced by CYP11B2 gene expression. Aldosterone will activate the mineralocorticoid receptor, leading to increased reabsorption of natrium and water in kidneys and promoting potassium excretion (**Figure 3**).

Besides the systemic RAAS, a local system found in many tissues, including the heart has been described. The local RAAS functions in a dual manner: independently and in correlation with systemic RAAS components. This second pathway of metabolisation present in the heart consists in ANG II synthesis by the endopeptidase chymase. Chymase is a serine protease found in a type of granulocytes called mast cells, which are located in the heart interstitium. Mast cells contain granules

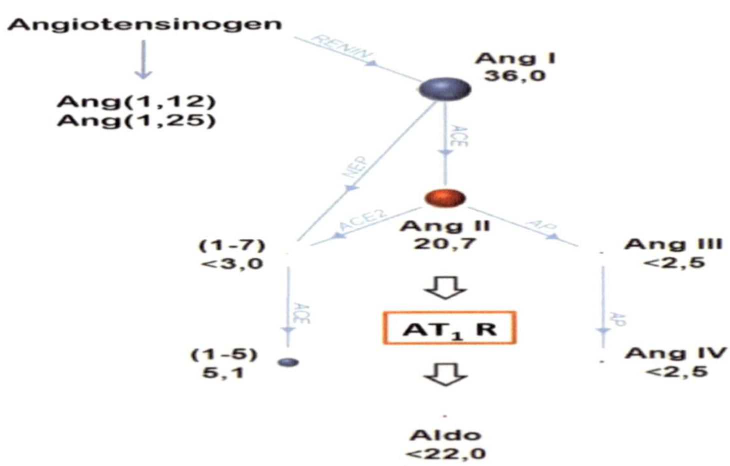

Figure 3.
The RAAS activation [13].

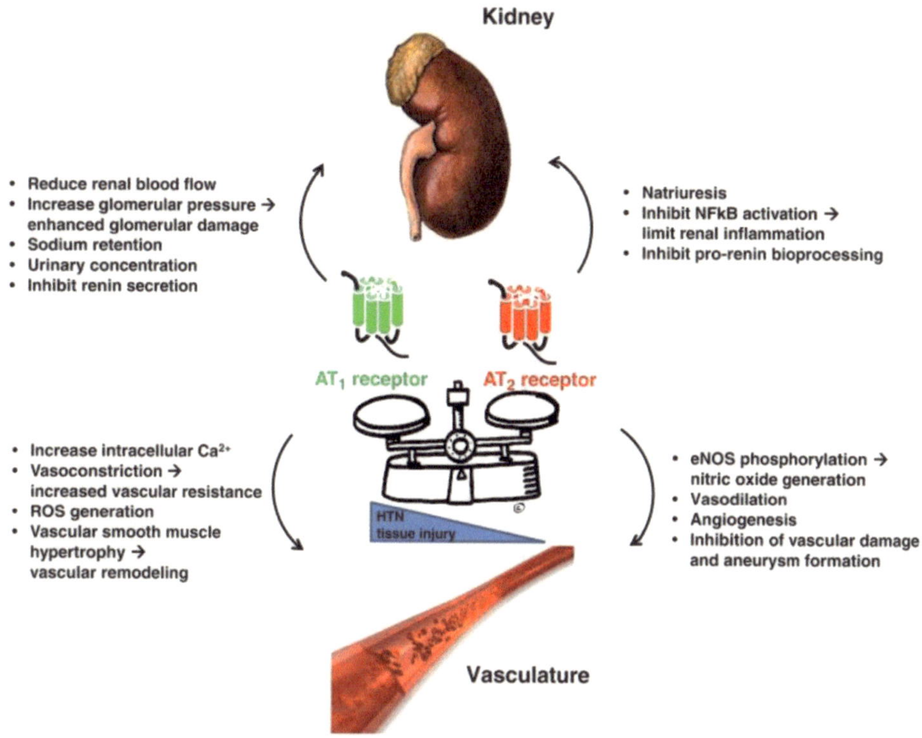

Figure 4.
The actions of angiotensin receptors [11].

that are filled with cytokines and proteases (including chymase). The degranulation releases chymase during the inflammatory process via a classically mediated ligand-dependent pathway. Chymase actions are similar to ACE but with a much higher catalytic activity in transforming ANG I into ANG II. Therefore, chymase may be in fact a major ANG II-forming enzyme in the human heart. Its role may be linked to many pathological processes, such as hypertension, atherosclerosis, vascular proliferation, development of cardiomyopathies, myocardial infarction and heart failure, as well as cardiac fibrosis [18–21].

A third pathway consists in the action of ACE2 (monocarboxypeptidase angiotensin-converting enzyme 2), which is similar to ACE in <50% of amino acidic constitution and therefore is unresponsive to the treatment with ACE inhibitors. This enzyme transforms ANG I into Ang 1-9 and ANG II into Ang 1-7, thus promoting the synthesis of cardiorenal-protective peptides, through diminishment of ANG I and ANG II levels. This pathway may lead to a key role in modulating ANG II actions in cardiovascular disease [22].

3. Gene involvement

3.1 ACE gene polymorphisms and heart disease

The implication of gene alterations in the ethiopathogeny of different diseases is an important study direction in the later years. The RAAS is not only involved in the cardiovascular homeostasy but also in the development of the coronary artery disease. It seems that angiotensin-converting enzyme gene mutations have the key role in this process.

ACE gene is found on the long arm of chromosome 17, position 23 (17q23), and include 26 exons and 25 introns (**Figure 5**) [23].

There are several possible mutations of this gene which initiate cardiovascular disease. The most common one is represented by deletions and insertions in the 16th intron (presence or absence of the 287-bp Alu repeat sequence), resulting in three possible genotypes, identified by the length of the fragments: II (490 bp), ID (490, 190 bp) and DD (190 bp) (**Figure 6**).

Mutations are determined by collecting 2 ml of blood from the patients, extracting the DNA from the leukocytes, followed by amplification of the genetic material, agarose gel electrophoresis in order to separate the fragments and UV light identification of coloured fragments according to their length.

Those genotypes interact with the conventional environmental factors and influence the severity of the coronary artery disease. Most of the studies have demonstrated that the D allele is a risk factor, while the I allele is a protective factor. This hypothesis was first formulated by Cambien et al., in the study published in Circulation in 1994 [24]. They postulated that D allele is correlated with high plasma levels of ACE and that genotypes ID and DD increase the predisposition for myocardial infarction. Since then, this theory was largely debated. Guney et al. performed a study on 203 patients with acute coronary syndrome or positive functional tests, selected based on the severity of the stenosis (>70%), as revealed by the coronary angiography. The control group was formed of 140 patients with nonsignificant coronary stenosis. Results showed a higher prevalence of the D allele in the patient group than the control group ($p = 0.002$), in patients with hyperlipidaemia ($p = 0.009$) and smoking habit ($p = 0.004$), and a higher number of diseased vessel (two/three vessel diseases) ($p = 0.002$). The conclusion drawn was that D allele interacts with conventional risk factor and determines the degree of severity, since patients with D allele have higher plasma levels of ACE and are therefore more exposed to angiotensin II.

Dhar et al., in a study on 217 patients with coronary disease compared to 255 control patients (with negative Treadmill test), also found the D allele as an independent risk factor in coronary vessel disease ($p < 0.05$). Most of the ID and DD patients were heavy smokers, diabetic and dyslipidemic, suggesting the interaction between D allele and risk factors. Another association pointed out by other studies is that D allele is more frequent in obese patients. The high levels of angiotensin II might lead to disturbances in macronutrient oxidation, causing fat deposits and weight gain [25, 26]. Despite the fact that most of the studies focusing on this mutation were positive, results are still controversial, since there are a few studies which showed negative results. Poorgholi et al. concluded from their 1050 CAD patients' study that I/D polymorphism is not a predisposant factor [27]. Fujimura et al. also found no association between ACE gene polymorphism and CAD in his study on 1840 patients (947 ischemic patients and 893 healthy controls) [28]. Although in some populations, there were no positive correlations between D allele and coronary disease; this may be due to geographical differences and ethnical differences. In general, D allele seems to be a predisposing factor, and probably the severity of the ischemic heart disease (infarction or unstable angina, one/two/three

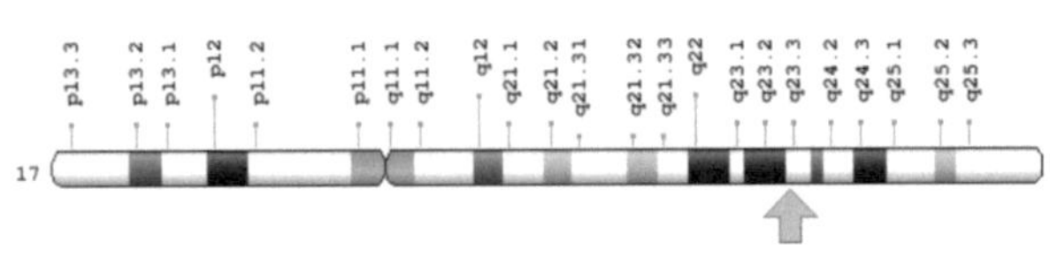

Figure 5.
The ACE gene location [29].

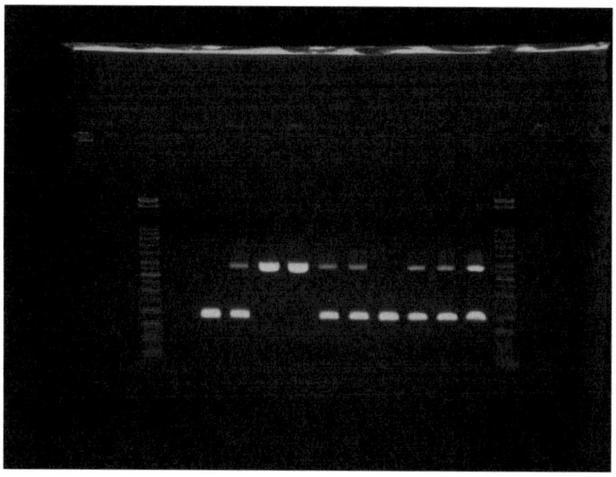

Figure 6.
ACE genotypes: II (490 bp), ID (490, 190 bp) and DD (190 bp) [30].

vessel diseases) depends on the interactions between this mutation and the other risk factors: hypertension, diabetes mellitus, obesity, smoking habit and dyslipidaemia. In most studies, patients with DD genotype were hospitalized for myocardial infarction and had multiple vessel disease and a high exposure to the conventional cardiovascular risk factors [31, 32]. The role of D allele in atherosclerosis was generally studied by measuring and analysing the carotid artery intima-media thickness (IMT). Results were statistically significant for high-risk patients, rather than low-risk patients. Therefore, D variant carriers have an increased risk of atherosclerosis, but only if they are also exposed to other genetic or environmental risk factors [33].

The association between ACE gene mutations and hypertension is explained by high plasmatic levels of ACE and a consequently increased quantity of ANG II in D-allele carriers. However, the predisposition for preeclampsia in D-allele women is less understood, and it is just a hypothesis still under debate. Preeclampsia implies high blood pressure and proteinuria after 20 weeks of pregnancy. It is the consequence of many risk factors, both genetic and environmental, and the RAAS may play a key role, since inactivating this system leads to a failure in achieving the hypervolemic status needed during pregnancy [34]. The studies on pregnant women are rare and often very small, making the available data on this issue very inconsistent. Some evidence suggests that it may come as a result of an abnormal regulation of the RAAS system and high intrauterine artery resistance [35]. Abedin et al., in a study on 296 patients, found no association between I/D polymorphisms but demonstrated a positive correlation with another ACE gene mutation, ACE rs4343, which was significantly present in the case group versus control group ($p = 0.0001$) [36]. On the other hand, a large meta-analysis including 45 studies found that DD and ID pregnant women have a higher risk of pregnancy-induced hypertension than II patients (74% D allele in the case group versus 56% D allele in the control group) [37]. Still, the large heterogeneity of the included patients lowers the force of the obtained results.

Another line of study in this field is represented by the involvement of RAAS system in the development of hypertrophic cardiomyopathy. This condition requires an anterior interventricular septum larger than 13 mm or a posterior interventricular septum larger than 15 mm, in the absence of hypertension and valvular disease. In most cases, hypertrophy is asymmetric. The implication of sarcomere protein gene mutations in the physiopathology of this disease is widely acknowledged. Later studies bring into attention a new puzzle piece in this genetic disorder, suggesting that ACE gene may also be involved, since the inhibition of ACE reduces cardiac hypertrophy and remodellation post-myocardial infarction [38, 39]. Some studies suggest that ACE gene mutations explain the intrafamilial phenotype differences in hypertrophic cardiomyopathy, which is considered a monogenic disease [40].

3.2 ACE polymorphism and treatment response

Genetic background often explains the different response to medication of patients with the same medical conditions.

"Sodium sensitive" hypertensive patients have certain physiopathological particularities: enhanced activity of the sympathetic nervous system, reduced renal excretion of sodium, hypersensitivity to vasoconstrictive hormones (adrenaline, ANG II) and alterations of the counteracting atrial natriuretic peptide (ANP) [41]. The connection between salt and hypertension is still under investigation, and it has not been fully explained [42], although it is clear that for some hypertensive patients, the mean blood pressure is higher when salt ingestion is increased. Studies focusing on determining the frequency of sodium sensitivity in hypertensive patients revealed the following distribution: 51% high sodium sensitivity, 33% intermediate sensitivity and 16% sodium resistance [43]. Patients with D allele display a higher sensitivity to dietary sodium intake, and therefore the risk of developing hypertension in those patients is increased, when high-sodium/low-potassium diet is associated. However, when the diet is low in sodium, the risk between presence and absence of D allele seems to be the same [44]. ACE inhibitors represent a gold standard therapy for cardiovascular and renal disease. But almost half of the hypertensive patients respond to ACE inhibitors alone. Most patients require multiple drug associations. Previous studies analysed the role of ACE gene polymorphisms in modulating treatment response in different populations around the world, and the results are still inconsistent. Heidari et al. in their study on 72 newly diagnosed hypertensive patients revealed a stronger response to enalapril and lisinopril in the DD genotype than ID ($p = 0.03$) and II ($p = 0.001$) [45]. Arnett et al. demonstrated a greater response to hydrochlorothiazide in hypertensive women during a 4-week study [46]. Results are still controversial. Other studies found no relationship between RAAS inhibitory treatment and genetic polymorphisms. For example, the study performed by Millions et al. did not link the D allele with a supplementary reduction of blood pressure during the treatment with ACE inhibitors [47]. Despite the results of some important trials stating that DD patients have a better response to ACE inhibitors, those results turn out non-reproducible [48]. Moreover, the pharmacogenomic analysis from PROGRESS study (5685 patients) based on the neuroprotective effect of perindopril after stroke revealed no implication of the ACE polymorphism on the end points [49]. There is also some evidence pointing out a different response to hydrochlorothiazide according to this polymorphism and sex: diuretic treatment was more efficient in II women and DD men [50]. Others suggested the synergism between II, ID and a-adducin Gly/Trp as an even better predictor of response to diuretic treatment [50]. The importance of the polymorphism in salt-sensitive population, the size of the study population and the ethnic variability may partially account for those differences. The differences in the response to ACE inhibitors between Caucasians and Africans are overwhelming [51], and this definitely represents a solid motivation for further investigations in this field. As for the response to beta blockers, a very important piece in the hypertension puzzle, it seems that ACE gene mutations might also be involved. A randomized controlled trial in hypertensive patients treated with atenolol demonstrated a more important blood pressure decrease in patients with an AGT M 235 T or G-6A genotype. But those results were not reproducible [48].

Pharmacogenomics is a new medical field trying to achieve personalized treatments according to our genetic package, in an attempt to obtain better results in the treatment-resistant patients. Given that RAAS is a key system in cardiovascular haemodynamic, discovering mutations that explain individual differences in efficacy and toxicity of various drug categories, as well as designing treatment strategies taking genetic data in consideration, will certainly be a revolutionary breakthrough in the future, both for hypertension and coronary artery disease.

4. Renin-angiotensin-aldosterone therapeutical blockade

The blockade of RAAS has become a central therapeutic strategy for patients with cardiac pathology. The inhibition of RAAS system is realized with modulating drugs such as ACEi, angiotensin receptor blockers (ARBs) and mineralocorticoid receptor antagonists (MRA) [52, 53].

The direct renin inhibition is obtained upstream of the action of the converting enzyme and has been well-known for decades. This type of blockade prevents the generation of ANG I and subsequently ANG II. The most used drug is named aliskiren; it has been evaluated in two clinical heart failure trials: ASTRONAUT and ATMOSPHERE. These trials proved that aliskiren failed to reduce cardiovascular death or heart failure hospitalization, but it induced a significantly more important decrease from baseline in NT-proBNP levels [54].

One of the latest advances in RAAS is represented by the development of innovative AT1 receptor blockers which have a dual action, going beyond simple antagonism of the binding of ANG II. Similar to the classic ARBs, these molecules will act on the superfamily of G-protein-coupled receptors (GPCRs). When GPCRs are activated by an agonist, they will determine intracellular dissociation of a heterotrimeric G protein into G and G subunits. The result consists in inducing second messenger-mediated cellular responses. Other groups of proteins which induce specific signalling pathways in a manner that is independent of the protein G are represented by the 13-arrestins [52, 54].

Even though efficient blockade of RAAS system can be obtained at different levels, the cornerstone therapy remains the use of angiotensin-converting enzyme inhibitors (ACEi), ARBs and MRA. RAAS blockade combined with natriuretic peptides augmentation is considered to be a revolutionary treatment in heart failure patients.

The renin-angiotensin-aldosterone system and the natriuretic peptides have a yin/yang relationship that has a great potential in inducing beneficial effects, by influencing both systems. The desired effects of the RAAS blockade can be enhanced by augmenting the natriuretic peptides activity. Even though the monotherapy with neprilysin inhibitors has failed to prove efficiency in heart failure patients, the association with a RAAS blocker has overcome the disappointing results, and even more, it has proven to be one of the most promising therapies in patients with different types of heart failure [55–57].

5. Conclusions

The renin-angiotensin-aldosterone system is a key piece in the puzzle of cardiovascular homeostasis and disease. Therapeutical targeting of this system at different levels represents an important medical breakthrough. The observation that patients with the same condition and exposure to risk factors may have different disease evolutions and responses to treatment led to genetic studies and to the discovery of the importance of ACE D allele in the determinism of coronary artery disease, arterial hypertension and hypertrophic cardiomyopathy. Therefore, developing treatment strategies according to the specific genetic pattern of the patient is the next step in medical evolution.

Conflict of interest

The authors declare no conflict of interest.

Author details

Manuela Ciocoiu, Iris Bararu-Bojan*, Maria Vladeanu* and Codruta Badescu
Department of Pathophysiology, University of Medicine and Pharmacy "Grigore T. Popa", Iasi, Romania

*Address all correspondence to: iris_bararu@yahoo.com and maria.apavaloaie@gmail.com

References

[1] Kanwar YS, Venkatachalam MA. Comprehensive Physiology. Ultrastructure of Glomerulus and Juxtaglomerular Apparatus. John Wiley & Sons, Inc; 2010

[2] Taugner R, Hackenthal E. The Juxtaglomerular Apparatus: Structure and Function. Heidelberg, Germany: Springer Verlag; 1989

[3] Lu H, Cassis LA, Kooi CW, Daugherty A. Structure and functions of angiotensinogen. Hypertension Research. 2016;**39**(7):492-500. DOI: 10.1038/hr.2016.17. [published correction appears in Hypertension Research 2016 Nov;39(11):827]

[4] Wu C, Lu H, Cassis LA, Daugherty A. Molecular and pathophysiological features of angiotensinogen: A mini review. American Journal of the Medical Sciences (Boston). 2011;**4**:183-190

[5] Streatfeild-James RM, Williamson D, Pike RN, Tewksbury D, Carrell RW, Coughlin PB. Angiotensinogen cleavage by renin: Importance of a structurally constrained N-terminus. FEBS Letters. 1998;**436**:267-270

[6] Zhou A, Carrell RW, Murphy MP, Wei Z, Yan Y, Stanley PL, et al. A redox switch in angiotensinogen modulates angiotensin release. Nature. 2010;**468**:108-111

[7] Inagami T, Iwai N, Sasaki K, Yamamo Y, Bardhan S, Chaki S, et al. Cloning, expression and regulation of angiotensin II receptors. Journal of Hypertension. 1992;**10**:713-716

[8] Murphy TJ, Alexander RW, Griendling KK, Runge MS, Bernstein KE. Isolation of a cDNA encoding the vascular type-1 angiotensin II receptor. Nature. 1991;**351**:233-236

[9] Munter K, Hackenthal E. The effects of endothelin on renovascular resistance and renin release. Journal of Hypertension. Supplement. 1989;7:S276-S277

[10] Nakamoto H, Ferrario CM, Fuller SB, Robaczewski DL, Winicov E, Dean RH. Angiotensin-(1-7) and nitric oxide interaction in renovascular hypertension. Hypertension. 1995;**25**:796-802

[11] Sparks MA, Crowley SD, Gurley SB, Mirotsou M, Coffman TM. Classical renin-angiotensin system in kidney physiology. Comprehensive Physiology. 2014;**4**(3):1201-1228. DOI: 10.1002/cphy.c130040

[12] Patel VB, Zhong JC, Grant MB, Oudit GY. Role of the ACE2/angiotensin 1-7 axis of the renin-angiotensin system in heart failure. Circulation Research. 2016;**118**(8):1313-1326. DOI: 10.1161/CIRCRESAHA.116.307708

[13] Ames MK, Atkins CE, Pitt B. The renin-angiotensin-aldosterone system and its suppression. Journal of Veterinary Internal Medicine. 2019;**33**(2):363-382. DOI: 10.1111/jvim.15454

[14] Swedberg K, Eneroth P, Kjekshus J, Wilhelmsen L. Hormones regulating cardiovascular function in patients with severe congestive heart failure and their relation to mortality. CONSENSUS Trial Study Group. Circulation. 1990;**82**(5):1730-1736

[15] Anker SD, Chua TP, Ponikowski P, et al. Hormonal changes and catabolic/anabolic imbalance in chronic heart failure and their importance for cardiac cachexia. Circulation. 1997;**96**(2):526-534

[16] Broqvist M, Dahlstrom U, Karlberg BE, et al. Neuroendocrine response in acute heart failure and the influence of treatment. European Heart Journal. 1989;**10**(12):1075-1083

[17] Levine TB, Francis GS, Goldsmith SR, Simon AB, Cohn JN. Activity of the sympathetic nervous system and renin-angiotensin system assessed by plasma hormone levels and their relation to hemodynamic abnormalities in congestive heart failure. The American Journal of Cardiology. 1982;**49**:1659-1666

[18] Froogh G, Pinto JT, Le Y, et al. Chymase-dependent production of angiotensin II: An old enzyme in old hearts. American Journal of Physiology. Heart and Circulatory Physiology. 2016;**312**(2):H223-H231. DOI: 10.1152/ajpheart.00534.2016

[19] Badir PM, Walston JD, Carey RM. Subcellular characteristics of functional intracellular renin-angiotensin systems. Peptides. 2012;**38**:437-445. DOI: 10.1016/j.peptides.2012.09.016

[20] Ahmad S, Simmons T, Varagic J, Moniwa N, Chappell MC, Ferrario CM. Chymase-dependent generation of angiotensin II from angiotensin-(1-12) in human atrial tissue. PLoS ONE. 2011;**6**:e28501. DOI: 10.1371/journal.pone.0028501

[21] Alzayadneh EM, Chappell MC. Angiotensin-(1-7) abolishes AGE-induced cellular hypertrophy and myofibroblast transformation via inhibition of ERK1/2. Cellular Signalling. 2014;**26**:3027-3035. DOI: 10.1016/j.cellsig.2014.09.010

[22] Ferrario CM, Mullick AE. Renin angiotensin aldosterone inhibition in the treatment of cardiovascular disease. Pharmacological Research. 2017;**125**(Pt A):57-71. DOI: 10.1016/j.phrs.2017.05.020

[23] Hubert C, Houot AM, Corvol P, et al. Structure of angiotensin I converting enzyme gene. Two alternate promoters correspond to evolutionary steps of duplicated gene. The Journal of Biological Chemistry. 1991;**266**:15377-15383

[24] Cambien F, Costerousse O, Tiret L, et al. Plasma level and gene polymorphism of angiotensin converting enzyme in relation to myocardial infarction. Circulation. 1994;**90**:669-676

[25] Guney AI, Ergec D, Kirac D, et al. Effects of ACE polymorphisms and other risk factors on the severity of the coronary artery disease. Genetics and Molecular Research. 2013;**12**(4):6895-6906

[26] Rigat B, Hubert C, Alhenc-Gelas F, et al. An insertion/deletion polymorphism in the angiotensin I-converting enzyme gene accounting for half the variance of serum enzyme levels. The Journal of Clinical Investigation. 1990;**86**:1343-1346

[27] Poorgholi L, Saffar H, Fathollahi SM, et al. Angiotensin-converting enzyme insertion/deletion polymorphism and its association with coronary artery disease in an Iranian population. Journal of Tehran University Heart Center. 2013;**8**(2):89-94

[28] Fujimura T, Yokota M, Kato S, et al. Lack of association of angiotensin converting enzyme gene polymorphism or serum enzyme activity with coronary artery disease in Japanese patients. American Journal of Hypertension. 1997;**10**:1384-1390

[29] https://ghr.nlm.nih.gov/gene/ACE

[30] Apăvăloaie M, Bararu I, Jitaru D, Ciocoiu M, Bădescu M, Arsenescu Georgescu C. Genetic determinism in the acute coronary syndrome. Romanian Journal of Artistic Creativity. 2016;**4**(10):173-181. ISSN 2327-5707, e ISSN 2473-6562

[31] Freitas AI, Mendonca I, Brion M. RAS gene polymorphisms, classical

risk factors and the advent of coronary artery disease in the Portuguese population. BMC Cardiovascular Disorders. 2013;**8**:15

[32] Niemec P, Zak I, Wita K. The risk of coronary artery disease associated with cigarette smoking and hypercholesterolemia is additionally increased by the presence of the AT1R gene 1166c allele. Biochemical Genetics. 2008;**46**:799-809

[33] Sayed-Tabatabaei FA, Houwing-Duistermaat JJ, van Duijn CM, Witteman JC. Angiotensin converting enzyme gene polymorphism and carotid artery wall-thickness: A meta-analysis. Stroke. 2003;**34**(7):1634-1639

[34] Rahimi Z, Rahimi Z, Mozafari H, Parsian S. Preeclamspia and angiotensin converting enzyme I/D and angiotensin II type-I receptor (ATIR) A1166C polymorphisms: Association with ACE I/D polymorphism. Journal of the Renin-Angiotensin-Aldosterone System. 2012;**14**(2):174-180

[35] Shah DM. The role of RAAS in the pathogenesis of preeclampsia. Current Hypertension Reports. 2006;**8**:144-152

[36] Abedin DA, Esmaeilzadeh E, Amin-Beidokhti M, et al. ACE-gene rs4343 polymorphism elevates the risk of preeclampsia in pregnant women. Journal of Human Hypertension. 2018;**32**:825-830

[37] Miao HW, Gong H. Correlation of ACE gene deletion/insertion polymorphism and risk of pregnancy-induced hypertension: A meta-analysis based on 10,236 subjects. Journal of the Renin-Angiotensin-Aldosterone System. 2015;**16**(4):982-994

[38] Linz W, Schaper J, Wiemer G, Albus U, Scholkens BA. Ramipril prevents left ventricular hypertrophy with myocardial fibrosis without blood pressure reduction: A one year study in rats. British Journal of Pharmacology. 1992;**107**:970-975

[39] Pfeffer MA, Braunwald E, on behalf of SAVE investigators. Effect of captopril on mortality and morbidity in patients with left ventricular dysfunction after myocardial infarction. New England Journal of Medicine. 1992;**327**:669-677

[40] Lechin M, Quinones M, Omnran A, et al. Angiotensin-I converting enzyme genotypes and left ventricular hypertrophy in patients with hypertrophic cardiomyopathy. Circulation. 1995;**92**:1808-1812

[41] Beeks E et al. Genetic predisposition to salt-sensitivity: A systematic review. Journal of Hypertension. 2004;**22**:1243-1249

[42] Lev-Ran A, Porta M. Salt and hypertension: A phylogenetic perspective. Diabetes/Metabolism Research and Reviews. 2005;**21**:118-131

[43] Campese VM. Salt sensitivity in hypertension. Renal and cardiovascular implications. Hypertension. 1994;**23**:531-550

[44] van der Kleij FG, de Jong PE, Henning RH, de Zeeuw D, Navis G. Enhanced responses of blood pressure, renal function, and aldosterone to angiotensin I in the DD genotype are blunted by low sodium intake. Journal of the American Society of Nephrology. 2002;**13**(4):1025-1033

[45] Heidari F, Vasudevan R, Ali SZ, Etemabad A. Association of insertion/deletion polymorphism of angiotensin-converting enzyme gene among Malay male hypertensive subjects in response to ACE inhibitors. Journal of the Renin-Angiotensin-Aldosterone System. 2015;**16**(4):872-879

[46] Arnett DK, Davis BR, Ford CE, et al. Pharmacogenetic association of

the angiotensin-converting enzyme insertion/deletion polymorphism on blood pressure and cardiovascular risk in relation to antihypertensive treatment: The genetics of hypertension-association treatment study. Circulation. 2005;**111**:3374-3383

[47] Millions HJ, Kostapanos MS, Vakalis K, et al. Impact of renin-angiotensin-aldosterone system genes on the treatment response of patients with hypertension and metabolic syndrome. Journal of the Renin-Angiotensin-Aldosterone System. 2007;**8**:181-189

[48] Mellen PB, Herrington DM. Pharmacogenomics of blood pressure response to antihypertensive treatment. Journal of Hypertension. 2005;**23**:1311-1325

[49] Harrap SB et al. The ACE gene I/D polymorphism is not associated with the blood pressure and cardiovascular benefits of ACE inhibition. Hypertension. 2003;**42**:297-303

[50] Schwartz GL et al. Interacting effects of gender and genotype on blood pressure response to hydrochlorothiazide. Kidney International. 2002;**62**:1718-1723

[51] Sciarrone MT et al. ACE and alpha-adducin polymorphism as markers of individual response to diuretic therapy. Hypertension. 2003;**41**:398-403

[52] von Lueder TG, Sangaralingham SJ, Wang BH, et al. Renin-angiotensin blockade combined with natriuretic peptide system augmentation: Novel therapeutic concepts to combat heart failure. Circulation. Heart Failure. 2013;**6**(3):594-605. DOI: 10.1161/CIRCHEARTFAILURE.112.000289

[53] Zannad F, McMurray JJ, Krum H, van Veldhuisen DJ, Swedberg K, Shi H, et al. Eplerenone in patients with systolic heart failure and mild symptoms. The New England Journal of Medicine. 2011;**364**:11-21

[54] Krum H, Massie B, Abraham WT, Dickstein K, Kober L, McMurray JJ, et al. Direct renin inhibition in addition to or as an alternative to angiotensin converting enzyme inhibition in patients with chronic systolic heart failure: Rationale and design of the Aliskiren Trial to Minimize OutcomeS in Patients with HEart failuRE (ATMOSPHERE) study. European Journal of Heart Failure. 2011;**13**:107-114

[55] Cleland JG, Swedberg K. Lack of efficacy of neutral endopeptidase inhibitor ecadotril in heart failure. The International Ecadotril Multi-centre Dose-ranging Study Investigators. Lancet. 1998;**351**:1657-1658

[56] Brenner BM, Cooper ME, de Zeeuw D, Keane WF, Mitch WE, Parving HH, et al. Effects of losartan on renal and cardiovascular outcomes in patients with type 2 diabetes and nephropathy. The New England Journal of Medicine. 2001;**345**:861-869

[57] Dahlof B. Left ventricular hypertrophy and angiotensin II antagonists. American Journal of Hypertension. 2001;**14**:174-182

Chapter 3

Significance of the Renin-Angiotensin System in Clinical Conditions

Vedran Đambić, Đorđe Pojatić, Anto Stažić and Aleksandar Kibel

Abstract

The renin-angiotensin system, in both its circulating and local tissue roles, is intertwined with multiple other regulatory and signalling mechanisms in various tissues and organ systems. It plays a central role in the normal regulation of arterial blood pressure and in the development of hypertension, which is an immense global public health burden and a crucial modifiable risk factor in the development of cardiovascular diseases. The renin-angiotensin system plays also important roles in a range of other clinical conditions such as heart failure, kidney failure, diabetes mellitus and others. Therapeutic interventions within the renin-angiotensin system include the use of medications such as angiotensin-converting enzyme inhibitors and angiotensin receptor antagonists, which are well established and have been invaluable as clinically effective tools during many years of practical use. Additionally, numerous other therapeutic approaches targeting components of the renin-angiotensin system have been developed or are currently in development. This chapter will discuss details of the roles of this system in the most relevant clinical conditions.

Keywords: renin-angiotensin system, hypertension, heart failure, angiotensin-converting enzyme inhibitors, diabetes mellitus

1. Introduction

The renin-angiotensin system (RAS), in both its circulating and local tissue roles, is intertwined with multiple other regulatory and signalling mechanisms in various tissues and organ systems. It is thereby involved in a number of clinical disorders with complex pathophysiological mechanisms. This chapter aims at providing a brief overview of the most relevant conditions in clinical practice in which the RAS is important when considering its pathophysiological or therapeutic significance.

2. RAS in hypertension

Systemic arterial hypertension is the most crucial global modifiable risk factor for morbidity and mortality and clearly connected to cardiovascular disease development [1]. The RAS plays a central role in the normal regulation of blood pressure and in the development of hypertension [2]. RAS regulates blood pressure through its effector protein angiotensin II (Ang II), which causes the constriction of the efferent arteriole in the kidney glomerulus, as well as the vasoconstriction of arterioles in peripheral circulation [3]. In addition to the above, RAS achieves its effects by stimulating the secretion of aldosterone by adrenal glands and vasopressin by the posterior pituitary, which, through the synergistic effect of water retention and sodium absorption, ensures the stability or increase of the intravascular volume [3]. Abnormal RAS activity is present in conditions such as hypertension and diabetes, which causes slow, continuous damage to the renal parenchyma and leads to the development of chronic kidney disease (CKD) [4, 5].

The pathophysiology of essential hypertension and other types of hypertension is extremely complex and not fully elucidated, with the RAS contributing only in part to the multifactorial pathophysiological mechanisms [1]. There is a pathophysiological interplay between the RAS, endothelial function/dysfunction, the role of the sympathetic nervous system, natriuretic peptides, inflammation and the immune system [1]. A dysfunction in the factors that contribute to blood pressure control may lead to the development of increases in mean blood pressure [1]. The genetic background/predisposition is of course very relevant. Furthermore, the RAS may play a completely different role depending on the type of hypertension. For instance, in primary aldosteronism, the most common type of secondary hypertension, there is an inappropriate increase in aldosterone synthesis in the setting of low plasma renin [6].

In addition to its main hemodynamic, fluid volume and vascular tone-changing effects, RAS can participate in stimulating immune cell infiltration, inflammation and fibrosis that is present in conditions such as renal ischemia, myocardial infarction and systemic hypertension [7]. Because of the connection of angiotensin II to endothelial dysfunction, proinflammatory and profibrotic actions and oxidative stress, it is associated with renal, cardiac and vascular injury and thereby directly linked to target organ damage in hypertension [7].

Besides the most largely know effector arm of the RAS, which includes angiotensin II and the angiotensin II type 1 receptor (AT1R), the RAS system is much more complex and features other important components, including angiotensin-(1–7) (Ang-(1–7), which exerts opposite effects than angiotensin II—it leads to vasodilation, and it has antifibrotic and antiproliferative effects [8]. Angiotensin-(1–7) is being extensively studied, including for its potential antihypertensive effects. Its potentially favourable cardiovascular actions are elicited through Mas G protein-coupled receptors that are expressed in various tissues essential to blood pressure regulation such as the brain, kidneys, blood vessels and heart [9]. However, the clinical potential and role of angiotensin-(1–7) in the pathophysiology and treatment is not yet clear because of a lack of adequate clinical studies and data [9].

From a therapeutic standpoint, there are several important groups of pharmacological agents targeting the RAS in hypertension. These include already clinically well-established antihypertensive drugs such as inhibitors of the angiotensin-converting enzyme (ACE inhibitors), AT1R blockers or direct renin inhibitors [2]. However, novel potential drugs targeting the RAS are also being developed, including agents seeking to upregulate the ACE2/angiotensin-(1–7)/Mas axis—counteracting the unwanted actions of the ACE/angiotensin II/AT1R axis—or agents such as new small molecule inhibitors, recombinant ACE2 protein, as well as gene therapy suppressing angiotensinogen at the RNA level [2].

3. Role in heart failure

3.1 Introduction

Heart failure is a progressive condition defined by the inability of the heart to pump enough blood to meet the body requirements for nutrients and oxygen.

It develops when the heart fails to pump enough blood to meet the requirements of metabolizing tissues, which is caused by conditions like ischaemic heart disease, arrhythmias, etc., and produces different symptoms and signs. Some of the symptoms and signs include the following: exertional dyspnoea and/or dyspnoea at rest, orthopnoea, acute pulmonary oedema, chest pain, tachycardia, fatigue and weakness, distention of neck veins, rales, wheezing, S gallop or pulsus alternans, etc. [7].

Heart failure is one of the most challenging pathophysiologic states with increasing prevalence in the Western world, affecting 1–2% of the total population [10].

The neurohumoral basis of heart failure is complex and includes interaction between components of the renin-angiotensin-aldosterone system (RAAS), the sympathetic nervous system and counterregulatory mediators (i.e., the natriuretic peptides).

Activation of neurohumoral mechanisms (i.e., the renin-angiotensin-aldosterone system) initially improves cardiac output, but despite this, counterregulatory mechanisms finally lead to heart failure syndrome with different signs and symptoms [11].

3.1.1 The neurohumoral pathophysiology of heart failure

The renin-angiotensin-aldosterone system (RAAS) plays a major role in the pathogenesis of heart failure, and the impact of treatment that targets RAAS is significant.

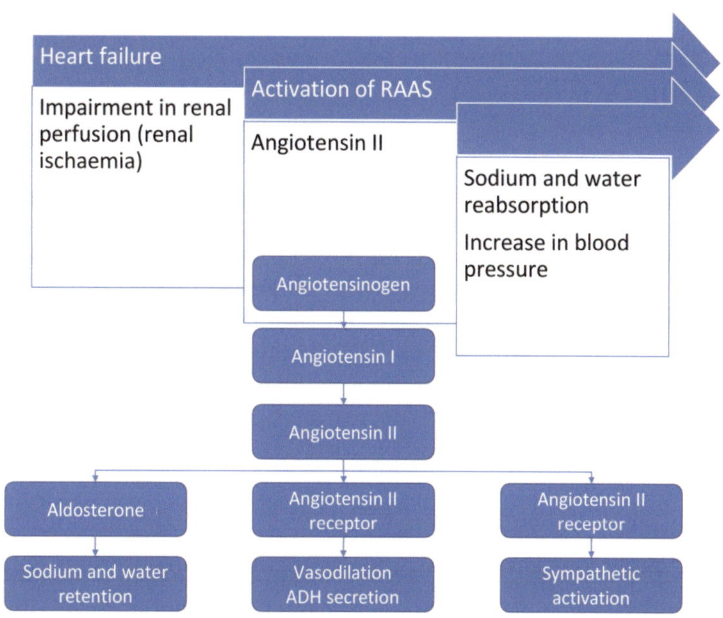

Figure 1.
A summary of the key pathogenesis of RAAS in heart failure. RAAS, renin-angiotensin-aldosterone system; ADH, antidiuretic hormone.

The neurohumoral basis of heart failure includes activation of RAAS by renal hypoperfusion (caused by hypotension or hypovolemia) and sympathetic activation [12].

Production of angiotensin II leads to vasoconstriction of arterioles and an increase in blood pressure, leads to antidiuretic hormone secretion, and stimulates the release of aldosterone and consequently salt and water retention [12] (**Figure 1**).

All of the above temporarily improve cardiac output but finally propagate the heart failure syndrome via overwhelming the opposing vasodilators and natriuretic mediators (i.e., A-type natriuretic peptide and B-type natriuretic peptide) which are mainly secreted form heart in response to the degree of ventricular stretch [13].

An imbalance between regulatory and counterregulatory mechanisms is the major pathophysiological target in heart failure treatment.

Therefore drugs that modulate RAAS (i.e., ACE inhibitors, beta-blockers, angiotensin receptor blockers and mineralocorticoid receptor antagonists) represent the core of treatment augmented by angiotensin receptor neprilysin inhibitors as a novel therapy [14].

4. Role in kidney failure

Chronic kidney disease (CKD) is a disease whose prevalence is constantly increasing in the elderly in the current, modern age [15]. CKD has multiple causes, where different organ dysfunctions lead to the reduction of the glomerular filtration rate, a disorder of excretory and endocrine function of the kidney and a significant diminishment of quality of life of such patients [16, 17]. End-stage renal failure, where glomerular filtration rate falls to less than 15 ml/min, requires treating the patient suffering from CKD with haemodialysis, which is a process in which blood is filtered through a semipermeable membrane, intended to replace the excretory function of the kidney [18]. Considering the high costs of treatment of patients with CKD, much research has been dedicated to studying the pathophysiological mechanisms that contribute to its development and exacerbation with the intention of developing new medications and procedures in an attempt to slow down the increase in the prevalence of patients with CKD. Abnormal activity of the renin-angiotensin system (RAS), regardless of the primary cause of CKD, has an immense value in the results of many studies [19].

In patients with CKD, RAS activity is increased in the central and peripheral circulation, which is indirectly confirmed by the fact that the activity and expression of the angiotensin-converting enzyme (ACE) linearly increases as the glomerular filtration rate decreases in patients with CKD [20]. Although ACE is the main enzyme that leads to RAS activation in patients with CKD by converting angiotensin I (Ang I) into angiotensin II, certain studies have shown that in particular types of acute renal failure, such as the one caused by aristolochic acid, there is a more significant RAS activation through chymases and a relatively mild ACE activation [21]. Chymase inhibition causes the slowing of the progression of fibrotic and inflammatory changes in the renal parenchyma, which makes it possible to talk about the synergistic effect of ACE and chymases in the activation of RAS effectors [21]. The role of chymase is also crucial for RAS activation in patients with stage five CKD treated with haemodialysis [22]. Considering the lower-than-normal levels of ACE and high chymase levels in such patients than those in patients with the same stage of CKD who are undergoing conservative treatment, it is clear that in such patients, the high levels of angiotensin in the renal parenchyma are a consequence of the increased chymase activity [22]. Increased chymase and angiotensin II activity leads to faster development of renal fibrosis in patients treated with haemodialysis than that in patients with the same stage of CKD who are not treated with dialysis [22].

According to new findings, nearly all RAS components may have a big effect on the development of CKD and its predisposing factors, such as hypertension. The activity of the prorenin receptors (PRR), which is found in the collecting duct cells of the kidney and increases the affinity of renin for angiotensinogen, is significant for RAS activation [23]. PRR blockade in animal models causes the decrease of RAS activity, which indirectly decreases levels of proteinuria, macroscopic signs of interstitial fibrosis and renal fibrosis. Although the increased activity of the systemic RAS plays a large role in the development of CKD, new findings place greater emphasis on the role of the local RAS in the renal and brain parenchyma and their synergistic effects. Prolonged activation of the local RAS in the renal parenchyma causes the production of inflammatory cytokines IL-6 and TNF-alpha, which lead to the infiltration of inflammatory cells into renal tissue, tissue damage and fibrosis [24] (**Table 1**).

Recent studies have identified the Wnt/β catenin signalling pathway as the main pathway for expression of RAS components in renal tissue, and its blockade in animal models of CKD was connected to significant clinical improvement and reduction of proteinuria and creatininaemia [25]. The Wnt/β catenin signalling pathway is the main pathway for RAS expression in the hypothalamic paraventricular nucleus which indirectly affects the renal tissue RAS through activation of the sympathetic nervous system [26]. Blocking the central Wnt/β catenin signalling pathway leads to the inhibition of expression of the renal RAS components and to the slowing of the progression of renal fibrosis; it is thus clear that in the tissue, local RAS is regulated by the central nervous system [5, 26]. Inhibition of RAS through intraventricular injection of losartan leads to a decrease in peripheral sympathetic nervous system activity and renal RAS activity, which potentially makes the sympathetic nervous system one of the possible pathways of communication between renal RAS and brain RAS [5]. In animal models of CKD, where a strong expression of RAS components was noticed after incubation of renal cells with inhibitors of the Wnt/β catenin signalling pathway, there is mainly a reduced expression of profibrotic factors in the renal parenchyma. In one such model, positive feedback was noticed between RAS and the Wnt/β catenin signalling pathway, where angiotensin II increases the expression of proteins of that signalling pathway, which is probably also the mechanism used by the central and circulating RAS to stimulate the activity of local RAS [25].

The role of the Wnt/β catenin signalling pathway in RAS activation and local fibrosis is also confirmed by research done on its inhibitors. The Klotho gene, which plays a significant role in slowing the process of ageing, encodes the protein which inhibits the components of the Wnt/β catenin signalling pathway [27]. In renal tissue of animal models of CKD, the Klotho level is inversely associated with the levels of RAS components, and the Klotho expression level showed a negative correlation with the degree of renal fibrosis [27]. Experimental studies on animal models that proved the effect of vitamin D on the decreased level of CKD and renal fibrosis measure an increased Klotho level, decreased activation of the Wnt/β catenin signalling pathway and low tissue RAS activity in in vitro conditions [28]. Multiple experimental studies have demonstrated the connection between the Wnt/β catenin signalling pathway and harmful RAS activation in CKD and renal fibrosis, which paves the way for new research with the aim of finding efficient medications that would slow the development and progression of CKD by affecting the above components [24, 26].

Local renal RAS may be activated by the effects of free radicals, which stimulate the expression of RAS components at the molecular level of signalling pathways. The above factor is one of the main mechanisms examined in harmful RAS

Component of RAS system whose activation or inhibition is involved in the development of CKD	Explanation of mechanism	References (author, year and title of study)
Prorenin receptor placed in collecting duct cells of kidney	PRR increases the affinity of prorenin and renin for angiotensin, and blockade of this receptor decreases activity of components of RAS and reduces laboratory markers of CKD	[23] (Fang et al., 2018, Role of (pro)renin receptor in albumin overload-induced nephropathy in rats)
Wnt/β catenin signalling pathway as the main pathway for expression of RAS components in kidney and brain tissue (hypothalamic paraventricular nucleus)	Wnt/β catenin signalling pathway is the main pathway of molecular expression of RAS components in brain and kidney tissue; local brain RAS and renal RAS communicate through the sympathetic nervous system	[26] (Zhou et al., 2019, The regulation effect of WNT-RAS signalling in hypothalamic paraventricular nucleus on renal fibrosis [25] (Xiao et al., 2019, Wnt/β-catenin regulates blood pressure and kidney injury in rats)
Angiotensin-converting enzyme in systemic circulation	Activity of ACE in systemic circulation increases with an increase of severity of CKD and in that way leads to further deterioration of kidney function	[20] (Wolke et al., 2017, Serum protease activity in chronic kidney disease patients: The GANI_MED renal cohort)
Chymase enzyme as an alternative pathway of activation of angiotensin II	Chymase, as an alternative pathway of RAS activation, converts angiotensin I to angiotensin II; blockade of this enzyme prevents CKD progression	[21] (Hsieh et al., 2017, Renal chymase-dependent pathway for angiotensin II formation mediated acute kidney injury in a mouse model of aristolochic acid I-induced acute nephropathy)
Neprilysin as an secondary enzymatic pathway for activation of angiotensin-(1–7)	Neprilysin, similar to ACE2 enzyme, converts angiotensin II to angiotensin-(1–7) which binds to Mas receptors and causes vasodilatory and renoprotective effects; prevents CKD	[31] (Domenig et al., 2016, Neprilysin is a mediator of alternative renin-angiotensin-system activation in the murine and human kidney)
Angiotensin receptor-associated protein (ATRAP), which is associated with AT1 receptor	ATRAP inhibits expression of AT1 receptor with circulating angiotensin II; it disables RAS activity in the development and progression of CKD	[32] (Kobayashi et al., 2017, An angiotensin II type 1 receptor binding molecule has a critical role in hypertension in a chronic kidney disease model)

Table 1.
RAS in the pathophysiology of renal failure.

activation in diabetic nephropathy, which is one of the leading causes of CKD [5]. The effect of free radicals on the development of CKD was demonstrated in studies examining strong antioxidants, such as melatonin. The application of melatonin in animal models of CKD leads to a reduced expression of all RAS components, decreased levels of markers of interstitial fibrosis and increased expression of antioxidant enzymes, such as superoxide dismutase [29]. The role of RAS in CKD has also been examined at the level of angiotensin receptors. Angiotensin II mainly produces its effect by activating the AT1 receptor, through which it causes vasoconstriction and has a proinflammatory and profibrotic effect on renal parenchyma. In addition to the AT1 receptor, there is also the AT2 receptor, which is presumed to have a vasodilatory and renoprotective effect. In experiments on animal models of CKD using resveratrol, which is a selective AT1 receptor antagonist and AT2 receptor agonist, increased expression of antioxidant enzymes was demonstrated, as was a reduced expression of fibronectin and type IV collagen—which are markers of tissue fibrosis [30].

A similar role, but with another agonist as the AT2 receptor, is played by the Mas receptor, which is also a part of RAS that contributes to local vasodilatation and has renoprotective effects. It is also known that, in normal circumstances, the Mas receptor is activated by angiotensin-(1–7), which is synthesized by the action of angiotensin-converting enzyme 2 (ACE2) on angiotensin II [31]. More recent studies of animal models with ACE2 enzyme blocking have determined the existence of stable levels of angiotensin-(1–7), and neprilysin was found to be the secondary enzymatic pathway contributing to its increase [31].

The AT1 receptor activity is regulated by a special associated protein, which regulates its expression and thus indirectly also regulates RAS activity. ATRAP inhibits the expression of the AT1 receptor and thus decreases the effect that circulating angiotensin II has on blood pressure levels, while also having a long-term effect by inhibiting the secretion of TNF-alpha, whose inflammatory factors contribute to the development of CKD [32].

The results of the latest studies confirm the essential role that RAS plays in the development and progression of CKD, which emphasizes the need and paves the way for new research on the subject of medications and therapeutic procedures that could efficiently decrease the growing epidemic of patients with CKD by affecting RAS components [32].

5. Role in diabetes

The RAS has been implicated in the human pathophysiology of numerous diseases, of particular importance in the development and onset of complications of diabetes as one of the most lethal non-communicable metabolic diseases [33]. The major intracellular mechanisms responsible for the adverse effects of diabetes are excessive production of advanced glycation end-products (AGEs), activation of the hexosamine biosynthetic pathway, activation of protein kinase C (PKC), lipotoxicity, mitochondrial dysfunction, enhanced oxidative stress and activation of intracellular RAS [34]. Because of hyperglycaemia in diabetes, intracellular accumulation of glucose that is not oxidised by glycolysis (because of inhibition of glyceraldehyde 3-phosphate dehydrogenase (GAPDH) by reactive oxygen species (ROS)) is already being diverted to other metabolic pathways [34]. The result of GAPDH inhibition is the increased flux of glucose metabolites through four other metabolic pathways. These are the aldose reductase or polyol pathway, the formation of advanced glycation end-products, the formation of diacylglycerol (DAG), resulting in protein kinase C activation and increased flux via the hexosamine biosynthesis pathway (HBP), and generation of the end product uridine diphosphate N-acetylglucosamine (UDP-GlcNAc), a substrate used for protein glycosylation [35]. The polyol pathway leads to NADPH and glutathione deficiency resulting in massive ROS formation [34]. On the other hand, in addition to leading to the development of insulin resistance, hyperlipidaemia also has a lipotoxic effect on cells [34]. Excessive fatty acid (FFA) inflow leads to increased and inefficient oxidation of FFA in the mitochondria with consequent generation of reactive oxygen radicals and exacerbation of mitochondrial dysfunction [34]. In diabetes, mitochondrial function decreases, and mitochondrial mass decreases, which is most likely connected to a decrease in OxPhos protein expression [34]. Excessive amounts of ROS, AGEs, DAG and activated PKCs and post-translational O-GlcNAcylation of transcription factors jointly lead to activation of intracellular synthesis of AGT, renin and chymase and thus activation of intracellular RAS. In addition, AGEs act

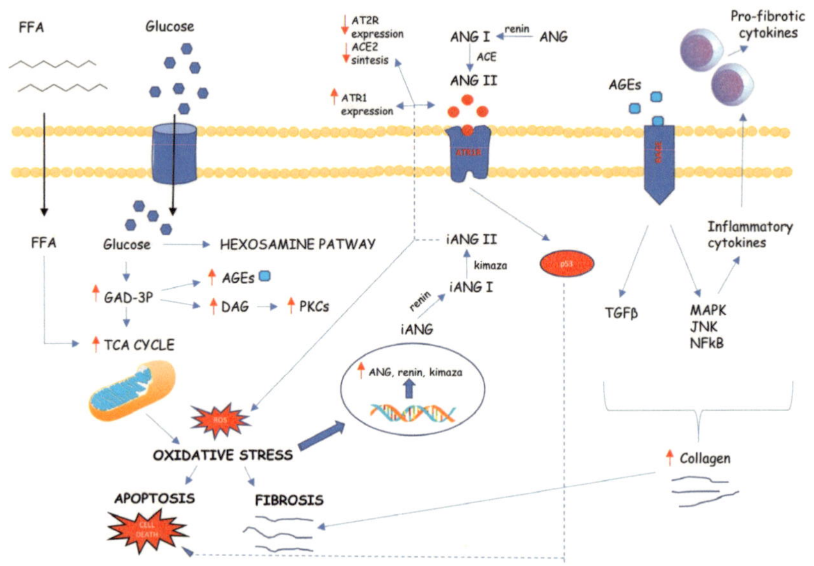

Figure 2.
Intracellular mechanism of activation and deleterious action of RAS in diabetes. ACE, angiotensin-converting enzyme; AGEs, advanced glycation end-products; Ang, angiotensinogen; Ang I, angiotensin I; Ang II, angiotensin II; ARB, angiotensin receptor blocker; AT1R, angiotensin II type 1 receptor; AT2R, angiotensin II type 2 receptor; DAG, diacylglycerol; FFA, free fatty acid; GAD-3P, glyceraldehyde 3-phosphate; iAng, intracellular angiotensin; JNK, c-Jun N-terminal kinase; MAPK, mitogen-activated protein kinase; NF-κB, nuclear factor-kappa beta; PKC, protein kinase C; RAGE, receptor for advanced glycation end-products; TCA cycle, tricarboxylic acid cycle; TGF-β, transforming growth factor beta.

via RAGE receptors that, through activation of inflammatory signalling pathways (TGF β, MAPK, JNK and NF-κB), lead to enhanced expression of collagen proteins and other extracellular matrix proteins. Together with activated RAS and activated PKCs isoforms by DAGs and by some lipid intermediates (e.g., ceramide, diacylglycerol and acylcarnitine), there is increased inflammation, fibrosis and apoptosis of cells attacked by hyperglycaemia [34] (**Figure 2**).

6. The role of intracellular Ang II (iAngII)

In addition to the commonly known circulating RAS, there is a local tissue intracellular RAS that occurs in all cells that produce peptides [36]. Activation of the transcription factors and gene expression described in the previous paragraph results in the synthesis of intracellular angiotensinogen, which is converted to angiotensin I by renin, and becomes the intracellular angiotensin II by the proteolytic action of chymase [34]. Chymase is a key proteolytic enzyme that completely replaces the role of extracellular angiotensin-converting enzyme within the cell and, unlike other alternative enzymes (tonin and cathepsin), has the highest substrate specificity. Intracellular Ang II levels are three times higher in diabetic patients than those in nondiabetic patients [37]. Intracellular Ang II has intracrine, autocrine, paracrine and endocrine functions [38]. iAngII by direct unidentified mechanisms, independent of global circulating RAS, causes an increase in oxidative stress and cell apoptosis and enhances the expression of intracellular components of RAS. Activation of AT1R via iAng II can trigger the MAPK signalling pathway (p38-MAPK), promoting inflammation, cell proliferation and thrombosis [37]. The study showed that an increase in phosphorylated p53 was associated with an increase in iAngII, enhanced activation of AT1R and RAS-independent O-glycosylation [39].

Also, iAng II enhances AT1R expression and decreases extracellular ACE2 expression in diabetic patients [37]. Increased expression of AT1R leads to stronger and negative effects of activation of AG1R [40]. ACE2 activation and AT2R activation are known to have protective effects on cells [38]. Direct inhibition, glycosylation or some other forms of protein modification leads to impaired ACE2 activity [41]. ACE2 catalyses the breakdown of Ang II to produce angiotensin-(1–7), which has an anti-inflammatory and antioxidant role [41]. In an animal model, it has been shown that oral administration of the recombinant-rich bacterium *Lactobacillus paracasei* (LP) can express and deliver human ACE2 to the circulation, which prevents the development of diabetic retinopathy and probably other forms of diabetes complication [41]. This potent protective effect of AT2R is prevented by the activation of AT1R by either iAngII or eAngII which results in the direct inhibition of AT2R activity or reduced expression of AT2R[38]. On the other hand, eAngII acts through AT1R and AG2R, and activation of a particular receptor depends on their balance regulated by the indicated intracellular mechanisms [38]. Thus, activation of AT1R by extracellular or intracellular Ang II leads to activation of several signalling pathways, activation of growth factor receptors, promotion of ROS synthesis and other apoptotic and fibrotic responses [37]. Patients with diabetes have 10 times higher levels of prorenin than renin than healthy subjects, and prorenin is thought to contribute more to the pathogenesis of diabetic complications [40]. Also, hyperglycaemia induces an increase in aldosterone by activation of AT1R and an increase in local Ang II whose pathophysiological effects in diabetes are unknown, but are thought to synergize with Ang II causing inflammation, apoptosis and fibrosis [42].

7. Influence of the RAS on pancreatic tissue, muscle tissue and adipose tissue

Chymase is found in all cells, and predominantly in mast cells, so we can also associate inflammation with the activation of RAS [43]. Activated intracellular RAS is thought to induce growth factor and cytokine synthesis by acting directly on other cells, which explains proteinuria in patients with diabetes with ACE inhibitors, resistance to antihypertensives in diabetics with hypertension and higher cardiovascular mortality in diabetics with hypertension [44]. Ang II reduces blood flow through beta Langerhans cells [45]. Ang II is thought to directly affect beta cell function and mass, promoting inflammation, oxidative stress and fibrosis [45]. All these effects finally lead to a decrease in insulin secretion and thus at least partially participate in the development of diabetes when it comes to beta cells [45]. Since muscle tissue accounts for about 80% of insulin-stimulated glucose disposal, the decrease in tissue perfusion caused by RAS directly contributes to the decreased glucose uptake into muscle cells [45]. On the other hand, it also leads to microvascular dysfunction, but all of this needs to be explored in more detail in the future [46]. Blockade of RAS causes overexpression of the GLUT4 transporter in skeletal muscle and accumulation of bradykinin that stimulates glucose uptake into skeletal muscle [47]. Long-term valsartan treatment reduced oxidative stress, NF-κB activation and TNF-alpha expression in skeletal muscle [48]. iAngII synthesized in adipose tissue (AT) leads to adipocyte hypertrophy, increased lipid synthesis and storage and inhibition of lipolysis, thereby modulating the lipid capacity of adipocytes to develop and worsen insulin resistance [49]. Also, Ang II promotes AT gene expression by enhancing the synthesis of proinflammatory adipokines and thus macrophage infiltration [49]. In female mice lacking the angiotensin II type 2 receptor (AT2R), decreased insulin sensitivity to adipose tissue leads to compensatory adiponectinaemia [50]. Animal model research has shown that RAS in diabetic rats leads to remodelling of the sympathetic nervous system, increased oxidative stress and increased norepinephrine levels in the myocardium itself [51]. In addition to its role in controlling electrolytes, blood pressure and vascular tone, RAS is involved in the development of inflammation, oxidative stress, metabolic syndrome, diabetes and its complications [49, 52–55].

8. Use of RAS inhibitors in patients with diabetes

Therapy with angiotensin-converting enzyme inhibitors (ACEi) and angiotensin receptor blockers (ARB) is a mainstay of treatment for diabetes mellitus (DM) because they prevent the development of diabetic complications [56]. Also, these drugs reduce the incidence of newly acquired DM II by 22% in the high-risk population by improving insulin sensitivity and insulin secretion [49]. Unfortunately, ARB and ACE inhibitors do not block iAngII, but only the autocrine and paracrine effects of extracellular Ang II, while the renin inhibitor aliskiren blocks extracellular and intracellular Ang II synthesis; therefore, aliskiren of all RAS inhibitors tested has the most potent effect in inhibiting fibrosis and apoptosis of cells exposed [49]. Insulin treatment showed no pronounced effect on the inhibition of iAngII synthesis [57]. Numerous meta-analyses state that many RAS blockers in use today will not have an additional benefit for diabetics over nondiabetics [38]. Today, attention is increasingly focused on the pharmacological activation of angiotensin II type 2 receptor and angiotensin-converting enzyme 2, which has been shown to reduce oxidative stress, inflammation, fibrosis and cell apoptosis in diabetics [57]. The protective effect has been demonstrated in combination therapy with activator neprilysin inhibitors (NEPi) with angiotensin-converting enzyme 2 [57]. This combination in diabetic rats significantly contributes to the inhibition of inflammatory, profibrotic and apoptotic signalling and thus prevents the development of diabetic cardiomyopathy and nephropathy [58]. Concomitant activation of the angiotensin II type 1 receptor (AT1) and activation of glucagon-like peptide-1 (GLP-1) receptor in insulin resistance leads to a decrease in oxidative damage to renal function and a significant decrease in albuminuria [59]. Urinary angiotensinogen may serve as an important marker of RAS activity, which may help in the decision to initiate treatment with RAS inhibitors and prevent the development and progression of complications in patients with diabetes [59, 60]. Numerous studies are based on uncontrolled studies of RAS inhibition; therefore, some studies discuss the beneficial effects on insulin sensitivity [61–65], and controversial studies report conflicting data [66–69]. Dosage and duration of treatment can be important factors that can create controversy, and in the future, they present us with a challenge in discovering the beneficial effects of inhibiting potent RAS [49].

9. Role in other clinical conditions

The local renin-angiotensin system in blood vessels of the brain parenchyma plays a significant role in regulating blood flow. The central nervous system RAS is separated from the systemic circulation by the blood-brain barrier, which enables the homeostasis of vasoconstriction and vasodilation of blood vessels in the brain, independent of blood pressure changes. Components of local RAS, such as renin, angiotensin II, ACE and AT1 receptor, participate in the cascade that results in vasoconstriction and anti-inflammatory effects, while activation of the AT2 receptor by angiotensin II and activation of the Mas receptor by angiotensin (1–7) cause vasodilation and increase local brain tissue perfusion; the above has been demonstrated in experimental animal models [70]. Reduced levels of angiotensin-converting enzyme 2, which produces vasoprotective angiotensin (1–7), have been found in clinical trials of patients with acute ischemic stroke; its levels normalise only in the post-acute phase of the stroke [71, 72]. As regards the above, the most sensational research conducted is the research on experimental animal models in which a reduction in the size of the ischemic penumbra after a stroke was found upon the use of the AT1R antagonist and the AT2R agonist; this represents a new treatment option for patients with ischemic stroke [71, 73]. Apart from the effect it has on regulating cerebral circulation, RAS also has a great effect on learning and memory acquisition through the activation of the AT2 receptor within the hippocampus and the cortex [74]. In animal models of dementia induced by scopolamine, simultaneous oral treatment of mice with ACE and ARB inhibitors slows cognitive decline [75]. It has been confirmed in clinical trials that ACE inhibitors and ARBs have an effect on slowing the progression of dementia caused by Alzheimer's disease, cerebrovascular disease and other diseases [70].

Local RAS in the retina, which is also part of the central nervous system, plays a significant role in the development and progression of diabetic retinopathy. The effect of angiotensin II on the production of free radicals and inhibition of antioxidant enzymes, which causes an increase in VEGF production, which in turn leads to the development of diabetic retinopathy by stimulating angiogenesis, has been confirmed in experimental animal models [76].

RAS plays a significant, though as yet insufficiently explored, role in the regulation of physiological processes in the lung parenchyma, which is evident in the fact that long-term treatment by ACE inhibitors and ARBs decreases the incidence of inflammatory, infectious and structural diseases of the lung parenchyma [77].

Coronavirus disease 2019 (COVID-19) caused a serious pandemic, with cardiovascular disease (10.5%) having the highest mortality rate, especially if over 70 years of age [78]. The hypothesis of an association between the use of ACE inhibitors and COVID-19 is based on the fact that the COVID-19 agent (also known as SARS-CoV-2) uses the SARS-CoV receptor angiotensin-converting enzyme 2 for entry into target alveolar epithelial cells causing pneumonia [79]. ACE2 is found in endothelial cells, type I and II alveolar cells of the lungs, enterocytes, basal epidermal cells of the skin and epithelium of the nose, oral mucosa and nasopharynx, and ACE2 levels decline with age and are lower in patients with chronic diseases [80]. Cardiovascular patients taking ACE inhibitors or AT blockers significantly increase mRNA expression and activity of cardiac ACE2 [81]. By binding SARS-CoV-2, ACE2 is depleted, which inhibits the protective axis and may cause imbalances between the ACE/Ang II/AT1R pathway and the ACE2/Ang-(1–7)/Mas receptor in RAS, which may result in an increase in ACE activity and exacerbation of acute severe pneumonia by an increase in proinflammatory factors [82]. Animal models have also shown that acute ACE-induced lung injury results in increased ACE activity and Ang II expression, whereas ACE2 and Ang- (1–7) activities are reduced. In patients undergoing ACEi therapy, there is an increase in renin whose cascade response is directed towards the formation of ACE2 as an important balancing factor [83]. The binding of SARS-CoV-2 may decrease ACE2 function, leading to increased neutrophil infiltration into the lungs and resulting in excessive inflammation and injury [84]. The progression of inflammation could lead to hypoxia-induced enhanced expression of renin synthesis, which closes the vicious cycle, and it is clear that hypertension is an important risk factor in patients with COVID-19 [85, 86]. A meta-analysis conducted by Caldeira et al. did not highlight the protective effect of ACEi and ARBs [87]. A 5-year study showed that ACE administration had a higher risk of pneumonia than ARB administration. Low-dose lisinopril in neurological dysphagia does not reduce pneumonia but increases mortality. Compared to calcium channel blockers, ACE inhibitors showed a lower mortality rate in patients with viral pneumonia [88]. Paradoxically, some authors report that chronic administration of ACE inhibitors or ARBs leads to an increase in ACE2 activity, which reduces the risk of infection because ACE2 dysregulation is triggered by the binding of a virus that increases the production of protective angiotensin-(1–7) [89]. This is based on the knowledge that ACEI/ARBs have shown an anti-inflammatory effect in the lungs, and it is suggested that they reduce the risk of pneumonia in elderly patients with hypertension with Parkinson's disease, in patients after CVI and in patients with chronic obstructive pulmonary disease (COPD) [90]. Patients with viral pneumonia who retained ACE and ARB had lower mortality rates and the need for intubation [91]. A major Chinese study has reported that ACE2 expression level is not crucial for the severity of COVID-19 infection but also plays an important role in the immune response and viral particle count and that, given pressure control, ACEi and ARBs can be used in patients with new types of coronavirus pneumonia to reduce pulmonary inflammation and reduce patient mortality rates [80]. No information is currently available on the number of hospitalized COVID-19 patients with hypotension, but it can be considered as an important limiting factor for the use of ACE inhibitors or ARBs in the treatment of COVID-19 patients [89]. Thus, we do not yet have the most accurate information on the real impact of ACEi and ARBs on infectious diseases such as pneumonia. Research should be conducted as soon as possible to prove the hypotheses and thus reduce the risk of mortality in patients taking ACE inhibitors and/or AT blockers.

At its molecular level, vitamin D is a prorenin molecule transcription antagonist. In animal models with knockout genes for vitamin D receptors, this is reflected in an elevated transcription of renin-angiotensin system components, development of hypertension and chronic cardiac and renal failure [92]. Data obtained through clinical studies that investigated possible therapeutic uses of vitamin D supplementations on RAS was inconsistent. Exogenous application of vitamin D has a distinct effect on local kidney tissue RAS. It causes a reduction in angiotensin excretion in patients with chronic renal failure and/or in renal transplant recipients [93, 94]. However, in prospective studies, the simultaneous application of vitamin D and drugs that affect RAS on patients with arterial hypertension or in normotensive patients did not result in a decrease in arterial pressure or the difference in RAS component concentrations [95, 96]. The exception with regard to the previous statements is the prospective study carried out by Wu et al., in which a decrease in blood pressure values was noted after 6 months of perioral administration of high doses of vitamin D, as well as a decrease in atherosclerotic plaque circumference and thickness on epicardial blood vessels of test subjects that underwent a coronarography [97]. The latest data points out an evident need for supplementary studies that must include a sufficient period for observation and a sufficiently large dose of vitamin D supplementations that would achieve a clearly notable effect on the observed variables.

Ghrelin, which is a peptide hormone secreted by the stomach, fulfils its physiological role by binding to target receptors in the central nervous system in which it triggers the feeling of hunger or the need for sleep by regulating hormone levels [98]. However, according to newly acquired data, the acylated form of this enzyme plays a significant, opposite role in the cardiovascular system, namely, in the area of the heart ventricle and aorta. Unlike RAS, which causes ischemic expansion and tissue degradation in cases of acute myocardial infarction, ghrelin has an opposite effect that is realized through its receptors. It causes a reduction in angiotensin-converting and proapoptotic enzyme expression, along with a reduction in oxidative stress [99].

10. Therapy targeting the RAS

Knowledge of the physiology of RAS and participation in the pathophysiology of cardiovascular and other diseases has led to the development of drugs that have an effect on numerous stages of the synthesis of RAS components and their action on known RAS receptors. Blocking of RAS is achieved by direct inhibitors of the enzyme responsible for the synthesis of RAS component (renin-angiotensin-converting enzyme) and inhibition of receptors through which they exert their effects (angiotensin receptor blockers) [100]. In recent years, new RAAS components have been discovered whose pathways have protective effects, which have enabled the development of new therapeutic targets [101]. One of the most impressive approaches in the study of novel therapeutic methods is the selective deletion of hepatic angiotensinogen [102].

10.1 Direct renin inhibitors

A low molecular weight non-peptide renin inhibitor that is well tolerated and has clinical utility is called aliskiren [103]. Clinical trials have demonstrated a strong antihypertensive and organoprotective (cardioprotective, renoprotective) role for the renin inhibitor [104, 105]. It has been shown to be very good at regulating blood pressure in hyperthyroid rats [106]. The use of aliskiren showed a decrease in the antidiuretic effect of renin [107]. This drug has a potent antifibrotic activity mediated by the reduction of oxidative stress and fibrogenic cytokines in all tissues [108].

10.2 ACE inhibitors

ACE is a multifunctional enzyme with numerous biological substrates. These classic drugs directly inhibit the vasoconstrictive and proliferative effects of the already known ACE/Ang II/AT1R axis and indirectly stimulate the production of the vasoprotective and antiproliferative peptide Ang-(1–7) [100].

10.3 Angiotensin I receptor antagonists/blockers (ARBs)

Selective AT1 receptor blockers inhibit the effects of Ang II on AT1R. The main effect is the inhibition of vasoconstriction, i.e., the reduction of peripheral vascular resistance, without significantly affecting the heart rate [109]. Administration of these drugs significantly reduces cardiovascular mortality, stroke, myocardial infarction and the onset and development of complications of diabetes [110]. The consequence of AT1R blockade is an increase in the secretion of renin, ACE and consequently Ang II [109].

10.4 Other therapeutic compounds

10.4.1 β1 blockers

Blocking the sympathetic β1 receptor in plasma cells of the dense macula results in a decrease in plasma renin levels. An effect on the reduction of Ang II has been demonstrated. β1 blockers also reduce the conversion of prorenin to renin [111].

10.4.2 Prorenin receptor antagonist/blocker

The newly discovered prorenin receptor blocker is a decoy peptide for the handle region of the prorenin average (HRP) [112]. It binds competitively to the non-proteolytic domain of prorenin, thereby preventing its activation [113].

Studies have shown that its administration reduces renal and cardiac impairment without affecting blood glucose levels in diabetic and antihypertensive rats [112]. Transgenic rat models with overexpression of the prorenin receptor showed massive proteinuria and glomerulosclerosis, but administration of HRP significantly suppressed the production of proteinuria and glomerulosclerosis, but without affecting the level of circulating Ang II [114]. Another study questions the use of HRP in which there was no reduction in target organ damage in hypertensive rats, nor did the blockade affect the expression of the prorenin receptor. It is thought that HRP cannot block extracellular signal-regulated kinases (ERKs) that induce prorenin and renin and that its positive effect depends on an undefined mechanism and not on the antagonism of the prorenin receptor. Clearly, this receptor plays a major role in the human body, but its cellular biology and its impact on the cardiovascular system need further investigation [112].

10.4.3 Chymase inhibitors

The chymase enzyme is only active in damaged tissues resulting from the activation of intracellular RAS [115]. The chymase can convert TGF-beta and MMP-9 precursors into their active forms that induce inflammation and fibrosis. To date, there are no specific results from human studies, but animal models have shown that inhibitors of chymase enzyme prevent vascular proliferation, myocardial fibrosis after cardiac infarction, development of diabetic complications, development of skin keloid and the occurrence of abdominal aortic aneurysm [116].

10.4.4 Angiotensin II receptor agonist

The first synthesized oral agonist was called Compound 21 (C21), which showed organoprotective, anti-inflammatory, antithrombotic, antifibrotic and antiapoptotic effects in an animal model. It has a very high affinity for AT2R and a low affinity for AT1R [103]. C21 lowers mean arterial pressure and improves ventricular function after myocardial infarction [117]. It significantly stimulates the growth of the hippocampal neurons, and its role mediated through AT2R in enhancing congenital functions is further investigated [118].

10.4.5 Activation angiotensin-converting enzyme 2 and Mas receptor agonist

Activation of the protective axis of RAS by ACE2, Ang-(1–7) and Mas receptors achieves beneficial vascular effects. The aim of this therapeutic target is to enhance the effect of ACE2, reduce the level of circulating Ang II and redirect the formation of Ang-(1–7), which by acting on AT2R and Mas receptors, achieves its protective effect. For the time being, the administration of human recombinant ACE2 is known to have an antihypertensive effect. ACE2 diminazene aceturate (DIZE) activator also reduces body weight and markers of adipogenesis and improves plasma lipid profile. They have a protective and pancreatic effect because it increases proliferation and decreases β-cell apoptosis and promotes glucose-stimulated insulin secretion. Overexpression or expression of ACE2 enhances basal and insulin-stimulated glucose uptake into cells, especially hepatocytes and adipocytes. Another tested ACE2 activator is 1-[[2-(dimethylamino) ethyl] amino]-4-(hydroxymethyl)-7-[[(4-methylphenyl) sulfonyl] oxy] -9H-xanthone-9 (XNT), which, through 2 weeks of continuous treatment, led to an improvement in endothelial function in hypertensive and diabetic rats by reducing oxidative stress. It also causes Mas receptor-mediated vasodilation [119]. In diabetic rats, it reduced pulmonary hypertension and significantly improved function and reduced myocardial fibrosis [120]. Therefore, in addition to the antihypertensive effect, the use of ACE2 activator leads to beneficial metabolic effects that reduce the possibility of diabetes and its complications [119]. Deficiency of Mas receptor genes or blockade of Mas receptors in male mice results in a metabolic syndrome that includes hyperlipidaemia, hyperglycaemia, hyperinsulinaemia, increased insulin resistance, increased glucose intolerance and adiposity [121]. The use of Mas agonists AVE0991 in animal models showed mild cardioprotective effects and an effect on lowering glucose and lipid levels in the blood [122]. Another Mas agonist of hydroxypropyl-p-cyclodextrin (HPβCD), in addition to increasing the half-life of Ang-(1–7), reduced the adverse effects and magnitude of myocardial infarction, myocardial hypertrophy and isoproterenol-induced heart damage. The remaining two agonists CGEM856 and CGEM857 cause vasodilation and cardioprotection [123]. In addition to Mas receptors, there are probably other receptors through which Ang-(1–7) exerts its effects, but this needs to be revealed in the future [119]. Virus-mediated ACE2 gene expression, i.e., gene therapy, has shown cardioprotective and antihypertensive effects in animal models. ACE2 overexpression by gene therapy is not limited to the heart but shows effects in brain and lung tissue where it exhibits antihypertensive effects [123].

10.4.6 Analogue of Ang-(1–7)

Although it is a very attractive therapeutic target, there are very limited studies today because of the short lifespan of this hormone. The clinical studies conducted focused on the intra-arterial or intravenous administration of infusions of the Ang-(1–7) analogue, which exerts its immediate vasodilatory effects. ACE inhibitors and angiotensin receptor antagonists, by their action, indirectly redirect the metabolic pathway of RAS towards the formation of increased levels of Ang-(1–7) [124].

10.4.7 Vaccines

The Ang I vaccine showed no antihypertensive effect, probably due to low antibody levels. Ang II vaccine, on the other hand, has proven to be very effective with a half-life of about 4 months. It was formed by covalent attachment of virus-like particles derived from the bacteriophage Qb envelope and modified Ang II. The antihypertensive effect manifested on systolic and diastolic pressures mostly in the early morning hours compared to placebo [125]. In addition to preclinical, it has also been tested in clinical trials with mild side effects such as flu symptoms. The more recent divalent vaccine HBcAg-CE12-CQ10 has shown antihypertensive and renoprotective properties without immune or electrophysiological adverse effects and has specific binding to AT1R and L-type calcium channels [126].

10.4.8 Alamandine

Recently, a new member of the RAS family named alamandine was discovered on animal models. It acts on MrgD receptors and exerts vasodilatory and cardioprotective effects similar to Ang-(1–7) [127]. Deletion of the MrgD gene at an early age resulted in the development of gender-independent dilated cardiomyopathy [128]. Endogenous alamandine is known to reduce leptin synthesis and secretion by activating the c-Src/p38 MAP kinase pathways, which is contrary to the action of Ang-(1–7) [129]. To date, no studies have been conducted on the administration of drugs that modulate this newly discovered component of RAS, so there is an opportunity to discover new therapeutic strategies in the future [127].

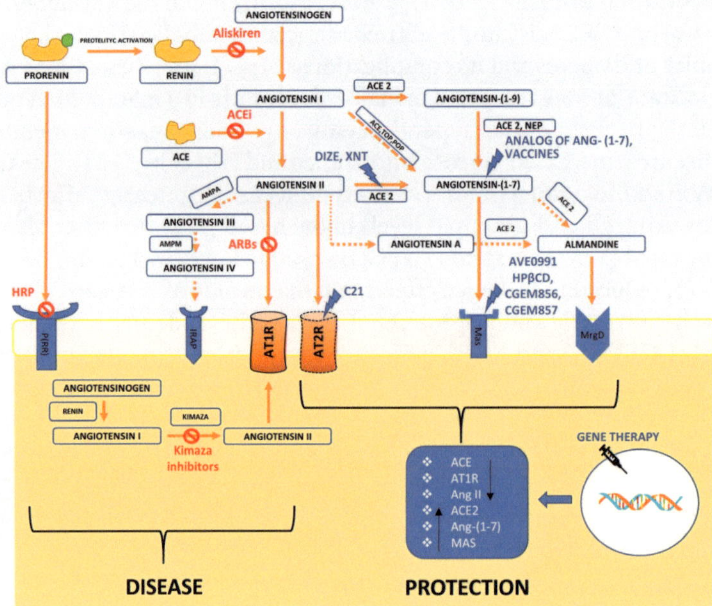

Figure 3.
Influence of specific therapeutic strategies on components of the RAS. ACE, angiotensin-converting enzyme; AMPA, aminopeptidase A; AMPM, aminopeptidase M; Ang, angiotensin; ARBs, angiotensin receptor blockers; AT1R, angiotensin II type 1 receptor; AT2R, angiotensin II type 2 receptor; AVE0991, orally active Mas receptor agonist; C21, Compound 21 (AT2R agonist); CGEM856 and CGEM857, orally active Mas receptor agonists; DIZE, ACE2 activator diminazene aceturate; HPβCD, Mas agonist hydroxypropyl-p-cyclodextrin; HRP, decoy peptide for handle region of the prorenin prosegment; IRAP, insulin-regulated aminopeptidase receptor; MasR, angiotensin-(1–7) Mas receptor; MrgD, Mas-related G protein-coupled receptor; NEP, neprilysin; POP, prolyl oligopeptidase; PRR, prorenin receptor; TOP, thimet oligopeptidase; XNT, ACE2 activator xanthenone.

10.5 Perspectives

The use of RAS-A blockers to date has not been shown to be sufficient to achieve the desired effects on the cardiovascular system [130]. With the realization that there is a protective axis within RAS, pharmacotherapy and gene therapy are increasingly focused on the activation of the protective components of RAS. The greatest challenge in the future remains to unravel the mutual mechanisms of regulation of the vasoconstrictive and vasoprotective axis. It is important to explore whether the expression of one enzyme/receptor induces the expression or inhibition of another, which are the mechanisms of action of the protective axis at the cellular level, which are the consequences of prolonged activation or inhibition of RAS, whether we can act more selectively on RAS and are there additional undetected roles and consequences of deleterious action therapies on RAS. We also need to further investigate the impact of gender, organ function, dosage and timing of administration on the targeted effect of therapy [123]. Discoveries and the complex modulation of RAS add a great challenge to the treatment of the most common diseases of humanity [130] (**Figure 3**).

Conflicts of interest

The authors have no conflict of interest to declare.

Author details

Vedran Đambić[1], Đorđe Pojatić[2,3], Anto Stažić[3,4] and Aleksandar Kibel[4,5]*

1 Department for Emergency Medical Services of the Osijek-Baranja County, Osijek, Croatia

2 Department of Internal Medicine, General Hospital Vinkovci, Vinkovci, Croatia

3 Faculty of Medicine Osijek, University J.J. Strossmayer in Osijek, Osijek, Croatia

4 Department for Heart and Vascular Diseases, Osijek University Hospital, Osijek, Croatia

5 Department of Physiology and Immunology, Faculty of Medicine, University J.J. Strossmayer in Osijek, Osijek, Croatia

*Address all correspondence to: aleksandar_mf@yahoo.com; alekibel@mefos.hr

References

[1] Oparil S, Acelajado MC, Bakris GL, Berlowitz DR, Cífková R, Dominiczak AF, et al. Hypertension. Nature Reviews. Disease Primers. 2018;**4**:18014. DOI: 10.1038/nrdp.2018.14

[2] Arendse LB, Danser AHJ, Poglitsch M, Touyz RM, Burnett JC Jr, Llorens-Cortes C, et al. Novel therapeutic approaches targeting the renin-angiotensin system and associated peptides in hypertension and heart failure. Pharmacological Reviews. 2019;**71**(4):539-570. DOI: 10.1124/pr.118.017129

[3] Mirabito Colafella KM, Bovée DM, Danser AHJ. The renin-angiotensin-aldosterone system and its therapeutic targets. Experimental Eye Research. 2019;**186**(May):107680

[4] Gant CM, Laverman GD, Vogt L, Slagman MCJ, Heerspink HJL, Waanders F, et al. Renoprotective RAAS inhibition does not affect the association between worse renal function and higher plasma aldosterone levels. BMC Nephrology. 2017;**18**(1):1-8

[5] Liu Y, Li L, Qiu M, Tan L, Zhang M, Li J, et al. Renal and cerebral ras interaction contributes to diabetic kidney disease. American Journal of Translational Research. 2019;**11**(5):2925-2939

[6] Byrd JB, Turcu AF, Auchus RJ. Primary aldosteronism. Circulation. 2018;**138**(8):823-835. DOI: 10.1161/CIRCULATIONAHA

[7] Ames MK, Atkins CE, Pitt B. The renin-angiotensin-aldosterone system and its suppression. Journal of Veterinary Internal Medicine. 2019;**33**(2):363-382. DOI: 10.1111/jvim.15454

[8] Kibel A, Novak S, Cosic A, Mihaljevic Z, Falck JR, Drenjancevic I. Hyperbaric oxygenation modulates vascular reactivity to angiotensin-(1-7) in diabetic rats: Potential role of epoxyeicosatrienoic acids. Diabetes & Vascular Disease Research. 2015;**12**(1):33-45. DOI: 10.1177/1479164114553424

[9] Medina D, Arnold AC. Angiotensin-(1-7): Translational avenues in cardiovascular control. American Journal of Hypertension. 2019;**32**(12):1133-1142. DOI: 10.1093/ajh/hpz146

[10] Mosterd A, Hoes AW. Clinical epidemiology of heart failure. Heart. 2007;**93**:1137-1146

[11] Broqvist M, Dahlstrom U, Karlberg BE, et al. Neuroendocrine response in acute heart failure and the influence of treatment. European Heart Journal. 1989;**10**(12):1075-1083

[12] Chaggar PS, Malkin CJ, Shaw SM, et al. Neuroendocrine effects on the heart and targets for therapeutic manipulation in heart failure. Cardiovascular Therapeutics. 2009;**27**:187-193

[13] McMurray JJ, Packer M, Desai AS, et al. Angiotensin-neprilysin inhibition versus enalapril in heart failure. The New England Journal of Medicine. 2014;**371**:993-1004

[14] Orsborne C, Chaggar PS, Shaw SM, et al. The renin-angiotensin-aldosterone system in heart failure for the non-specialist: The past, the present and the future. Postgraduate Medical Journal. 2017;**93**:29-37

[15] Karatas A, Canakci E, Turkmen E. Sociodemographic characteristics in patients with chronic kidney disease. Nigerian Journal of Clinical Practice. 2018;**21**(11):1461-1467

[16] Afsar B, Elsurer R. Association between serum bicarbonate and pH with depression, cognition and sleep quality in hemodialysis patients. Renal Failure. 2015;**37**(6):1-4

[17] He S, Zhu J, Jiang W, Ma J, Li G, He Y. Sleep disturbance, negative affect and health-related quality of life in patients with maintenance hemodialysis. Psychology, Health & Medicine. 2019;**24**(3):294-304

[18] Hrvačević R. Savremene metode dijalize-knjiga. 2012. pp. 1-227

[19] Murakami T, Iwamoto T, Yasuda G, Taniguchi M, Fujiwara A, Hirawa N, et al. Role of renin angiotensin system inhibitors in retardation of progression of end-stage renal failure: A retrospective study. Clinical and Experimental Nephrology. 2016;**20**(4):603-610

[20] Wolke C, Teumer A, Endlich K, Endlich N, Rettig R, Stracke S, et al. Serum protease activity in chronic kidney disease patients: The GANI_MED renal cohort. Experimental Biology and Medicine. 2017;**242**(5):554-563

[21] Hsieh WY, Chang TH, Chang HF, Chuang WH, Lu LC, Yang CW, et al. Renal chymase-dependent pathway for angiotensin II formation mediated acute kidney injury in a mouse model of aristolochic acid I-induced acute nephropathy. PLoS One. 2019;**14**(1):1-15

[22] Ohashi N, Isobe S, Ishigaki S, Suzuki T, Ono M, Fujikura T, et al. Intrarenal renin-angiotensin system activity is augmented after initiation of dialysis. Hypertension Research. 2017;**40**(4):364-370

[23] Fang H, Deng M, Zhang L, Lu A, Su J, Xu C, et al. Role of (pro)renin receptor in albumin overload-induced nephropathy in rats. American Journal of Physiology—Renal Physiology. 2018;**315**(6):1759-1768

[24] Bian X, Bai Y, Su X, Zhao G, Sun G, Li D. Knockdown of periostin attenuates 5/6 nephrectomy induced intrarenal renin–angiotensin system activation, fibrosis, and inflammation in rats. Journal of Cellular Physiology. 2019;**234**(12):22857-22873

[25] Xiao L, Xu B, Zhou L, Tan RJ, Zhou D, Fu H, et al. Wnt/β-catenin regulates blood pressure and kidney injury in rats. Biochimica et Biophysica Acta, Molecular Basis of Disease. 2019;**1865**(6):1313-1322

[26] Zhou G, Li J, Zeng T, Yang P, Li A. The regulation effect of WNT-RAS signaling in hypothalamic paraventricular nucleus on renal fibrosis. Journal of Nephrology. Apr 2020;**33**(2):289-297

[27] Zhou L, Mo H, Miao J, Zhou D, Tan RJ, Hou F, et al. Klotho ameliorates kidney injury and fibrosis and normalizes blood pressure by targeting the renin-angiotensin system. The American Journal of Pathology. 2015;**185**(12):3211-3223

[28] Eltablawy N, Ashour H, Rashed LA, Hamza WM. Vitamin D protection from rat diabetic nephropathy is partly mediated through Klotho expression and renin–angiotensin inhibition. Archives of Physiology and Biochemistry. 2018;**124**(5):461-467

[29] Ishigaki S, Ohashi N, Matsuyama T, Isobe S, Tsuji N, Iwakura T, et al. Melatonin ameliorates intrarenal renin–angiotensin system in a 5/6 nephrectomy rat model. Clinical and Experimental Nephrology. 2018;**22**(3):53949

[30] Jang IA, Kim EN, Lim JH, Kim MY, Ban TH, Yoon HE, et al. Effects of resveratrol on the renin-angiotensin system in the aging kidney. Nutrients. 2018;**10**(11):1-15

[31] Domenig O, Manzel A, Grobe N, Königshausen E, Kaltenecker CC, Kovarik JJ, et al. Neprilysin is a mediator of alternative renin-angiotensin-system activation in the murine and human kidney. Scientific Reports. 2016;**6**(August):1-11

[32] Kobayashi R, Wakui H, Azushima K, Uneda K, Haku S, Ohki K, et al. An angiotensin II type 1 receptor binding molecule has a critical role in hypertension in a chronic kidney disease model. Kidney International. 2017;**91**(5):1115-1125

[33] Connelly KA, Boyle AJ, Kelly DJ. Angiotensin II and the cardiac complications of diabetes mellitus. Current Pharmaceutical Design. 2007;**13**(26):2721-2729

[34] Zamora M, Villena JA. Contribution of impaired insulin signaling to the pathogenesis of diabetic cardiomyopathy. International Journal of Molecular Science. 2019;**20**(11):2833. DOI: 10.3390/ijms20112833

[35] Schaefer CF. Pathway databases. Annals of the New York Academy of Sciences. 2004;**1020**:77-91

[36] Bader M, Ganten D. Update on tissue renin-angiotensin systems. Journal of Molecular Medicine (Berlin, Germany). 2008;**86**:615-621

[37] Mandavia CH, Aroor AR, Demarco VG, Sowers JR. Molecular and metabolic mechanisms of cardiac dysfunction in diabetes. Life Sciences. 2013;**92**(11):601-608

[38] Singh VP, Le B, Bhat VB, Baker KM, Kumar R. High-glucose-induced regulation of intracellular ANG II synthesis and nuclear redistribution in cardiac myocytes. American Journal of Physiology. Heart and Circulatory Physiology. 2007;**293**(2):H939-H948

[39] Fiordaliso F, Leri A, Cesselli D, Limana F, Safai B, et al. Hyperglycemia activates p53 and p53-regulated genes leading to myocyte cell death. Diabetes. 2001;**50**(10):2363-2375

[40] Bernardi S, Michelli A, Zuolo G, Candido R, Fabris B. Update on RAAS modulation for the treatment of diabetic cardiovascular disease. Journal of Diabetes Research. 2016;**2016**:8917578. DOI: 10.1155/2016/8917578

[41] Verma A, Xu K, Du T, et al. Expression of human ACE2 in *Lactobacillus* and beneficial effects in diabetic retinopathy in mice. Molecular Therapy—Methods & Clinical Development. 2019;**14**:161-170

[42] Giacchetti G, Sechi LA, Rilli S, Carey RM. The renin-angiotensin-aldosterone system, glucose metabolism and diabetes. Trends in Endocrinology and Metabolism, 126. 2005;**16**(3):120

[43] Huynh K, Bernardo BC, McMullen JR, Ritchie RH. Diabetic cardiomyopathy: Mechanisms and new treatment strategies targeting antioxidant signaling pathways. Pharmacology & Therapeutics. 2014;**142**(3):375-415

[44] Barrett EJ, Eggleston EM, Inyard AC, Wang H, Li G, Chai W, et al. The vascular actions of insulin control its delivery to muscle and regulate the rate-limiting step in skeletal muscle insulin action. Diabetologia. 2009;**52**:752-764

[45] Chu KY, Lau T, Carlsson PO, Leung PS. Angiotensin II type 1 receptor blockade improves beta-cell function and glucose tolerance in a mouse model of type 2 diabetes. Diabetes. 2006;**55**:367-374

[46] Clerk LH, Vincent MA, Jahn LA, Liu Z, Lindner JR, Barrett EJ. Obesity blunts insulin-mediated microvascular

recruitment in human forearm muscle. Diabetes. 2006;**55**:1436-1442

[47] Miyata T, Taguchi T, Uehara M, Isami S, Kishikawa H, Kaneko K, et al. Bradykinin potentiates insulin-stimulated glucose uptake and enhances insulin signal through the bradykinin B2 receptor in dog skeletal muscle and rat L6 myoblasts. European Journal of Endocrinology. 1998;**138**:344-352

[48] Wei Y, Sowers JR, Clark SE, Li W, Ferrario CM, Stump CS. Angiotensin II-induced skeletal muscle insulin resistance mediated by NF-kappaB activation via NADPH oxidase. American Journal of Physiology. Endocrinology and Metabolism. 2008;**294**:E345-E351

[49] Goossens GH. The renin-angiotensin system in the pathophysiology of type 2 diabetes. Obesity Facts. 2012;**5**(4):611-624

[50] Quiroga DT, Miquet JG, Gonzalez L, Sotelo AI, Muñoz MC, et al. Mice lacking angiotensin type 2 receptor exhibit a sex-specific attenuation of insulin sensitivity. Molecular and Cellular Endocrinology. 2019;**498**:110587

[51] Yehya YM, Hussein AM, Ezam K, Eid EA, Ibraheim EM, et al. Blockade of renin angiotensin system ameliorates the cardiac arrythmias and sympathetic neural remodeling in hearts of type 2 DM rat model. Endocrine, Metabolic & Immune Disorders Drug Targets. 2019;**20**:464-478

[52] Sjølie AK. Prospects for angiotensin receptor blockers in diabetic retinopathy. Diabetes Research and Clinical Practice. 2007;**76**(Suppl 1): S31-S39

[53] Ghattas A, Lip PL, Lip GY. Renin-angiotensin blockade in diabetic retinopathy. International Journal of Clinical Practice. 2011;**65**:113-116

[54] Fletcher EL, Phipps JA, Ward MM, Vessey KA, Wilkinson-Berka JL. The renin-angiotensin system in retinal health and disease: Its influence on neurons, glia and the vasculature. Progress in Retinal and Eye Research. 2010;**29**:284-311

[55] Kalupahana NS, Moustaid-Moussa N. The renin-angiotensin system: A link between obesity, inflammation and insulin resistance. Obesity Reviews. 2012;**13**:136-149

[56] Mark PB, Papworth R, Ramparsad N, Tomlinson LA, Sawhney S, et al. Risk factors associated with biochemically detected and hospitalised acute kidney injury in patients prescribed renin angiotensin system inhibitors. British Journal of Clinical Pharmacology. Jan 2019;**86**:121-131

[57] Sharma N, Malek V, Mulay SR, Gaikwad AB. Angiotensin II type 2 receptor and angiotensin-converting enzyme 2 mediate ischemic renal injury in diabetic and non-diabetic rats. Life Sciences. 2019;**235**:116796

[58] Malek V, Sharma N, Gaikwad AB. Simultaneous inhibition of neprilysin and activation of ACE2 prevented diabetic cardiomyopathy. Pharmacological Reports. 2019; **71**(5):958-967

[59] Rodriguez R, Escobedo B, Lee AY, Thorwald M, Godoy-Lugo JA, et al. Simultaneous angiotensin receptor blockade and glucagon-like peptide-1 receptor activation ameliorate albuminuria in obese insulin-resistant rats. Clinical and Experimental Pharmacology & Physiology. Mar 2020;**47**(3):422-431

[60] Ba Aqeel S, Ye M, Wysocki J, et al. Urinary angiotensinogen antedates the development of stage 3 CKD in patients with type 1 diabetes

mellitus. Physiological Reports. 2019;**7**(19):e14242

[61] Hanley AJ, Zinman B, Sheridan P, Yusuf S, Gerstein HC. Effect of rosiglitazone and ramipril on β-cell function in people with impaired glucose tolerance or impaired fasting glucose: The DREAM trial. Diabetes Care. 2010;**33**:608-613

[62] Morel Y, Gadient A, Keller U, Vadas L, Golay A. Insulin sensitivity in obese hypertensive dyslipidemic patients treated with enalapril or atenolol. Journal of Cardiovascular Pharmacology. 1995;**26**:306-311

[63] Lender D, Arauz-Pacheco C, Breen L, Mora-Mora P, Ramirez LC, Raskin P. A double blind comparison of the effects of amlodipine and enalapril on insulin sensitivity in hypertensive patients. American Journal of Hypertension. 1999;**12**:298-303

[64] Shamiss A, Carroll J, Peleg E, Grossman E, Rosenthal T. The effect of enalapril with and without hydrochlorothiazide on insulin sensitivity and other metabolic abnormalities of hypertensive patients with NIDDM. American Journal of Hypertension. 1995;**8**:276-281

[65] Suzuki M, Ikebuchi M, Yokota C, Shinozaki K, Harano Y. Normalization of insulin resistance in non-obese essential hypertension by cilazapril treatment. Clinical and Experimental Hypertension. 1995;**17**:1257-1268

[66] Heinemann L, Heise T, Ampudia J, Sawicki P, Sindelka G, Brunner G, et al. Four week administration of an ACE inhibitor and a cardioselective beta-blocker in healthy volunteers: No influence on insulin sensitivity. European Journal of Clinical Investigation. 1995;**25**:595-600

[67] Petrie JR, Morris AD, Ueda S, Small M, Donnelly R, Connell JM, et al. Trandolapril does not improve insulin sensitivity in patients with hypertension and type 2 diabetes: A double-blind, placebo-controlled crossover trial. The Journal of Clinical Endocrinology and Metabolism. 2000;**85**:1882-1889

[68] Pratt MC, Lewis-Barned NJ, Walker RJ. A comparison between enalapril and captopril on insulin sensitivity in normotensive healthy volunteers. Australian and New Zealand Journal of Medicine. 1993;**23**:652-655

[69] Reaven GM, Clinkingbeard C, Jeppesen J, Maheux P, Pei D, Foote J, et al. Comparison of the hemodynamic and metabolic effects of low-dose hydrochlorothiazide and lisinopril treatment in obese patients with high blood pressure. American Journal of Hypertension. 1995;**8**:461-466

[70] Fouda AY, Artham S, El-Remessy AB, Fagan SC. Renin-angiotensin system as a potential therapeutic target in stroke and retinopathy: Experimental and clinical evidence. Clinical Science. 2016;**130**(4):221-238

[71] Bennion DM, Rosado CA, Haltigan EA, Regenhardt RW, Sumners C, Waters MF. Serum activity of angiotensin converting enzyme 2 is decreased in patients with acute ischemic stroke. Journal of the Renin-Angiotensin-Aldosterone System. 2016;**17**(3):1-7

[72] Kozak A, Ergul A, El-remessy AB, Johnson MH, Machado LS, Elewa HF, et al. Candesartan augments ischemia-induced proangiogenic state and results in sustained improvement after stroke. Stroke. 2009;**40**(5):18707

[73] Alhusban A et al. Compound 21 is pro-angiogenic in the brain and results in sustained recovery after ischemic stroke. Journal of Hypertension. 2015;**33**(1):170-180

[74] Jackson L, Eldahshan W, Fagan SC, Ergul A. Within the brain: The renin angiotensin system. International Journal of Molecular Sciences. 2018;**19**(3):1-23

[75] Ababei DC, Bild V, Ciobică A, Lefter RM, Rusu RN, Bild W. A comparative study on the memory enhancing actions of oral renin-angiotensin system altering drugs in scopolamine treated mice. American Journal of Alzheimer's Disease and Other Dementias. 2019;**34**(5):329-336

[76] Miller AG, Tan G, Binger KJ, Pickering RJ, Thomas MC, Nagaraj RH, et al. Candesartan attenuates diabetic retinal vascular pathology by restoring glyoxalase-I function. Diabetes. 2010;**59**(12):3208-3215

[77] Soto M, Bang SI, McCombs J, Rodgers KE. Renin angiotensin system-modifying therapies are associated with improved pulmonary health. Clinical Diabetes and Endocrinology. 2017;**3**(1):1-9

[78] The Novel Coronavirus Pneumonia Emergency Response Epidemiology Team. The epidemiological characteristics of an outbreak of 2019 novel coronavirus diseases (COVID-19) —China, 2020. China CDC Weekly. 2020;**2**(8):113-122

[79] Hoffmann M et al. The new 2019 coronavirus (2019-nCoV) uses the SARS coronavirus receptor ACE2 and the cellular protease TMPRSS2 to enter target cells. bioRxiv. 2020. 2020.01.31.929042

[80] Santos RAS, Sampaio WO, Alzamora AC, Motta-Santos D, Alenina N, et al. The ACE2/angiotensin-(1-7)/MAS axis of the renin-angiotensin system: Focus on angiotensin-(1-7). Physiological Reviews. 2018;**98**(1):505-553

[81] Zheng Y-Y, Ma Y-T, Zhang J-Y, Xie X. COVID-19 and the cardiovascular system. Nature Reviews Cardiology. May 2020;**17**(5):259-260

[82] Sun ML, Yang JM, Sun YP, Su GH. Inhibitors of RAS might be a good choice for the therapy of COVID-19 pneumonia. Zhonghua Jie He He Hu Xi Za Zhi. 2020;**43**(0):E014

[83] Ferrario CM, Jessup J, Chappell MC, Averill DB, Brosnihan KB, Tallant EA, et al. Effect of angiotensin-converting enzyme inhibition and angiotensin II receptor blockers on cardiac angiotensin-converting enzyme 2. Circulation. 2005;**111**(20):2605-2610

[84] Sodhi CP, Wohlford-Lenane C, Yamaguchi Y, Prindle T, Fulton WB, Wang S, et al. Attenuation of pulmonary ACE2 activity impairs inactivation of des-Arg9 bradykinin/BKB1R axis and facilitates LPS-induced neutrophil infiltration. American Journal of Physiology. Lung Cellular and Molecular Physiology. 2018;**314**(1):L17-L31

[85] Krämer BK et al. Effects of hypoxia on renin secretion and renal gene expression. Kidney International. 1998;**54**(67):S155-S158

[86] A top coronavirus doctor in Wuhan says high blood pressure is a major risk of death. Bloomberg News 9:00 PM GMT on 09 March 2020

[87] Caldeira D, Alarcão J, Vaz-Carneiro A, Costa J. Risk of pneumonia associated with use of angiotensin converting enzyme inhibitors and angiotensin receptor blockers: Systematic review and meta-analysis. BMJ. 2012;**345**:e4260. DOI: 10.1136/bmj.e4260

[88] Arai T, Yasuda Y, Takaya T, Toshima S, Kashiki Y, Yoshimi N, et al. ACE inhibitors and reduction of the risk of pneumonia in elderly people. American Journal of Hypertension. 2000;**13**(9):1050-1051

[89] Gurwitz D. Angiotensin receptor blockers as tentative SARS-CoV-2 therapeutics. Drug Developmental Research. 2020:1-4

[90] Kang JH, Kao LT, Lin HC, Wang TJ, Yang TY. Do outpatient statins and ACEIs/ARBs have synergistic effects in reducing the risk of pneumonia? A population-based case-control study. PLoS One. 2018;**13**(6):e0199981

[91] Henry C, Zaizafoun M, Stock E, Ghamande S, Arroliga AC, White HD. Impact of angiotensin-converting enzyme inhibitors and statins on viral pneumonia. Proceedings (Baylor University. Medical Center). 2018;**31**(4):419-423

[92] Legarth C, Grimm D, Wehland M, Bauer J, Krüger M. The impact of vitamin D in the treatment of essential hypertension. International Journal of Molecular Sciences. 2018;**19**(2):1-14

[93] Tiryaki O, Usalan C, Tarakcioglu M, Coban S. Calcitriol reduces albuminuria and urinary angiotensinogen level in renal transplant recipients. Transplantation Proceedings. 2018;**50**(5):1342-1347

[94] Tiryaki Ö, Usalan C, Sayiner ZA. Vitamin D receptor activation with calcitriol for reducing urinary angiotensinogen in patients with type 2 diabetic chronic kidney disease. Renal Failure. 2016;**38**(2):222-227

[95] McMullan CJ, Borgi L, Curhan GC, Fisher N, Forman JP. The effect of vitamin D on renin angiotensin system activation and blood pressure: A randomized control trial. Journal of Hypertension. 2017;**35**(4):822-829

[96] Zaheer S, Taquechel K, Brown JM, Adler GK, Williams JS, Vaidya A. A randomized intervention study to evaluate the effect of calcitriol therapy on the renin-angiotensin system in diabetes. Journal of the Renin-Angiotensin-Aldosterone System. 2018;**19**(1):1-8

[97] Wu Z, Wang T, Zhu S, Li L. Effects of vitamin D supplementation as an adjuvant therapy in coronary artery disease patients. Scandinavian Cardiovascular Journal. 2016;**50**(1):9-16

[98] Oberbach A, Schlichting N, Kullnick Y, Heinrich M, Lehmann S. Gastric mucosal devitalization improves blood pressure, renin and cardiovascular lipid deposition in a rat model of obesity. Endoscopy International Open. 2019;**7**(12):1605-1615

[99] Eid RA, El-Kott AF, Samir M, Zaki A, Eldeen MA, Al-hashem FH, et al. Acylated ghrelin protects aorta damage post-MI via activation of eNOS and inhibition of angiotensin-converting enzyme induced activation of NAD(P) H-dependent oxidase. Ultrastructural Pathology. 2018;**42**(5):416-429

[100] Atlas SA. The renin-angiotensin aldosterone system: Pathophysiological role and pharmacologic inhibition. Journal of Managed Care Pharmacy. 2007;**13**:S9-S20

[101] Ravandi A, Teo KK. Blocking the renin-angiotensin system: Dual-versus mono-therapy. Expert Review of Cardiovascular Therapy. 2009;**7**(6): 667-674. DOI: 10.1586/erc.09.47

[102] Lu H, Cassis LA, Kooi CWV, Daugherty A. Structure and functions of angiotensinogen. Hypertension Research. 2016;**39**:492-500

[103] Steckelings UM, Rompe F, Kaschina E, et al. The past, present and future of angiotensin II type 2 receptor stimulation. Journal of the Renin-Angiotensin-Aldosterone System. 2010;**11**:67-73

[104] Parving HH, Persson F, Lewis JB, Lewis EJ, Hollenberg NK, Avoid Study

Investigators. Aliskiren combined with losartan in type 2 diabetes and nephropathy. The New England Journal of Medicine. 2008;**358**:2433-2446

[105] McMurray JJ, Pitt B, Latini R, et al. Effects of the oral direct renin inhibitor aliskiren in patients with symptomatic heart failure. Circulation. Heart Failure. 2008;**1**:17-24

[106] Farhadi SAS, Dizaye KF. Aliskiren, fosinopril, and their outcome on renin-angiotensin-aldosterone system (RAAS) in rats with thyroid dysfunction. International Journal of Endocrinology. 2019;**2019**:5960563

[107] Yamashita S, Biswas KB, Nabi AHMN, Nakagawa T, Suzuki F, Ebihara A. Aliskiren reduces the release of soluble (pro)renin receptor from human umbilical vein endothelial cells. Biomedical Reports. 2018;**9**(3):247-252

[108] Altarejo Marin T, Machado Bertassoli B, Alves de Siqueira de Carvalho A, Feder D. The use of aliskiren as an antifibrotic drug in experimental models: A systematic review. Drug Development Research. 2019;**81**(1):114-126

[109] Brenner BM, Cooper ME, de Zeeuw D, et al. Effects of losartan on renal and cardiovascular outcomes in patients with type 2 diabetes and nephropathy. The New England Journal of Medicine. 2001;**345**:861-869

[110] Dahlof B, Devereux RB, Kjeldsen SE, et al. Cardiovascular morbidity and mortality in the Losartan Intervention For Endpoint reduction in hypertension study (LIFE): A randomised trial against atenolol. Lancet. 2002;**359**:995-1003

[111] Blumenfeld JD, Sealey JE, Mann SJ, et al. β-Adrenergic receptor blockade as a therapeutic approach for suppressing the renin-angiotensin-aldosterone system in normotensive and hypertensive subjects. American Journal of Hypertension. 1999;**12**:451-459

[112] Muller DN, Klanke B, Feldt S, Cordasic N, Hartner A, et al. (Pro) renin receptor peptide inhibitor "handle-region" peptide does not affect hypertensive nephrosclerosis in Goldblatt rats. Hypertension. 2008;**51**:676-681

[113] Ichihara A, Hayashi M, Kaneshiro Y, Suzuki F, Nakagawa T, Tada Y, et al. Inhibition of diabetic nephropathy by a decoy peptide corresponding to the "handle" region for nonproteolytic activation of prorenin. The Journal of Clinical Investigation. 2004;**114**:1128-1135

[114] Kaneshiro Y, Ichihara A, Sakoda M, Takemitsu T, Nabi AH, Uddin MN, et al. Slowly progressive, angiotensin II-independent glomerulosclerosis in human (pro)renin receptor-transgenic rats. Journal of the American Society of Nephrology. 2007;**18**:1789-1795

[115] Miyazaki M, Takai S, Jin D, Muramatsu M. Pathological roles of angiotensin II produced by mast cell chymase and the effects of chymase inhibition in animal models. Pharmacology & Therapeutics. 2006;**112**:668-676

[116] Takai S, Jin D, Miyazaki M. New approaches to blockade of the renin-angiotensin-aldosterone system: Chymase as an important target to prevent organ damage. Journal of Pharmacological Sciences. 2010;**113**:301-309

[117] Wan Y, Wallinder C, Plouffe B, Beaudry H, Mahalingam AK, Wu X, et al. Design, synthesis, and biological evaluation of the first selective nonpeptide AT2 receptor agonist. Journal of Medicinal Chemistry. 2004;**47**:5995-6008. DOI: 10.1021/jm049715t

[118] Li JM, Mogi M, Tsukuda K, Tomochika H, Iwanami J, Min LJ, et al. Angiotensin II-induced neural differentiation via angiotensin II type 2 (AT2) receptor-MMS2 cascade involving interaction between AT2 receptor-interacting protein and Src homology 2 domain-containing protein-tyrosine phosphatase 1. Molecular Endocrinology. 2007;**21**:499-511. DOI: 10.1210/me.2007-0111

[119] Fraga-Silva RA, Costa-Fraga FP, Murça TM, et al. Angiotensin-converting enzyme 2 activation improves endothelial function. Hypertension. 2013;**61**(6):1233-1238

[120] Ferreira AJ, Shenoy V, Qi Y, Fraga-Silva RA, Santos RAS, Katovich MJ, et al. Angiotensin-converting enzyme 2 activation protects against hypertension-induced cardiac fibrosis involving extracellular signal-regulated kinases. Experimental Physiology. 2011;**96**:287-294

[121] Santos SH, Fernandes LR, Mario EG, Ferreira AV, Pôrto LC, Alvarez-Leite JI, et al. Mas deficiency in FVB/N mice produces marked changes in lipid and glycemic metabolism. Diabetes. 2008;**57**(2):340-347

[122] Ebermann L, Spillmann F, Sidiropoulos M, Escher F, Heringer-Walther S, Schultheiss HP, et al. The angiotensin-(1-7) receptor agonist AVE0991 is cardioprotective in diabetic rats. European Journal of Pharmacology. 2008;**590**(1-3):276-280. DOI: 10.1016/j.ejphar.2008.05.024

[123] Ferreira AJ, Santos RA, Bradford CN, et al. Therapeutic implications of the vasoprotective axis of the renin-angiotensin system in cardiovascular diseases. Hypertension. 2010;**55**(2):207-213

[124] Schindler C, Bramlage P, Kirch W, Ferrario CM. Role of the vasodilator peptide angiotensin-(1-7) in cardiovascular drug therapy. Vascular Health and Risk Management. 2007;**3**(1):125-137

[125] Maurer P, Bachmann MF. Immunization against angiotensins for the treatment of hypertension. Clinical Immunology. 2010;**134**:89-95

[126] Wu H, Wang Y, Wang G, Qiu Z, Hu X, et al. A bivalent antihypertensive vaccine targeting L-type calcium channel and angiotensin II type 1 receptor. British Journal of Pharmacology. 2020;**177**:402-419

[127] Villela DC, Passos-Silva DG, Santos RA. Alamandine: A new member of the angiotensin family. Current Opinion in Nephrology and Hypertension. 2014;**23**(2):130-134. DOI: 10.1097/01.mnh.0000441052.44406.92

[128] Oliveira AC, Melo MB, Motta-Santos D, Peluso AA, Souza-Neto F, et al. Genetic deletion of the alamandine receptor MRGD leads to dilated cardiomyopathy in mice. American Journal of Physiology. Heart and Circulatory Physiology. 2019;**316**(1):H123-H133. DOI: 10.1152/ajpheart.00075.2018

[129] Uchiyama T, Okajima F, Mogi C, Tobo A, Tomono S, Sato K. Alamandine reduces leptin expression through the c-Src/p38 MAP kinase pathway in adipose tissue. PLoS One. 2017;**12**(6):e0178769. DOI: 10.1371/journal.pone.0178769

[130] Schmieder RE, Hilgers KF, Schlaich MP, Schmidt BM. Renin-angiotensin system and cardiovascular risk. Lancet. 2007;**369**:1208-1219

Chapter 4

Treatment of Essential Hypertension with Emphasis in the Renin-Angiotensin System: How to Prevent Secondary Outcomes without Adding Fuel to the Fire

Gabriel Lucca de Oliveira Salvador

Abstract

The effectiveness of angiotensin-converting enzyme inhibitors (ACEIs) and angiotensin II receptor blocker AT-1 (ARBs) in reducing the systemic hypertension (SH) is widely known. However their comparative outcomes resulting from prolonged use remain unknown. The objective of this chapter is to discuss the evidence of prospective randomized double-blind clinical trials; all the events result from prolonged use of ACEIs or ARBs in hypertensive patients. In lowering blood pressure, the use of ACE inhibitors or ARBs reduces, in long-term use, the risk of acute myocardial infarction, stroke, and heart failure. However, the use of ACEIs is effective in an overall quantitative analysis; the total mortality regarding cardiovascular causes an outcome that was not observed with the use of ARBs. This fact is assumed to be related to the higher plasma concentration of bradykinin in the use of ACEIs, a well-known cardiovascular-protective factor.

Keywords: hypertension, outcomes, ACEI, ARB, angiotensin-converting enzyme inhibitor, angiotensin II receptor blocker

1. Introduction

Hypertension is a manifestation in which there is an elevation of blood pressure levels to the point of causing imbalance in homeostasis and, consequently, leading the organism to a debilitated and pathological state with clinical repercussions [1]. Also, systemic arterial hypertension (SAH) is defined as "a multifactorial clinical condition characterized by elevated and sustained levels of blood pressure" [2, 3], considered when it is observed that a value >140 mmHg for systolic blood pressure and/or >90 mmHg for diastolic blood pressure [4].

SAH presents itself as one of the main risk factors for cardiovascular diseases [5], and in a recent study, it was shown as the risk factor with the highest number of citations in studies [6]; its incidence is increasing annually, reaching values of population of approximately 20% of Brazilians [7] or 24.3% [8]. Therefore, the control and prevention of SAH are fundamental to improve the quality of life and health indicators of the population.

The treatment of this clinical condition is based on three primordial objectives: the decrease in blood pressure levels; the maintenance of the desired levels, in view of the singular conditions of each patient; and adherence of the patient to the treatment as well as its continuity. From these perspectives, the measures in order to achieve this triad are based on two therapeutic pathways: changes in lifestyle and/or the use of drugs. Regarding changes in lifestyle, which also show efficacy in the prevention of hypertensive status, the most important are [9, 10] regular practice of physical activities, abstinence from alcohol, changes in food style prioritizing the decrease in salt intake and the increase in the consumption of fruit and vegetables, cessation of smoking, weight loss, control of psychosocial stress, slow breathing, monitoring, and surveillance by a multidisciplinary team.

Regarding the use of drug therapies, and according to their current availability, there are the following classes of medications [9]: diuretics, adrenergic inhibitors (central alpha-2 agonists, beta-blockers, and alpha-blockers), calcium channel blockers, angiotensin-converting enzyme inhibitors (ACEIs), angiotensin II receptor blockers AT1 (ARBs), and direct vasodilators such as nitrates.

Because they act primarily in one of the most important systems for the development of SAH, the renin-angiotensin-aldosterone system (SRAA) [11], four classes of medications stand out and are effective options in reducing morbidity and mortality due to cardiovascular diseases [12], direct renin inhibitors, ACEIs, ARBs, and aldosterone antagonists. Among these classes, the ACEIs and the ARBs are among the most prescribed options worldwide [11]. Therefore, the study and detailed knowledge of both classes is of fundamental importance.

The control of SAH is performed by the ACEI by blocking the transformation of angiotensin I in angiotensin II in the blood and tissues, while the ARB antagonizes the action of angiotensin II by means of the specific blockade of their AT1 receptors. In addition, studies show that the use of these antihypertensive patients is closely related to the decrease in the progression of renal diseases and in the prevention of heart and/or vascular diseases, such as acute myocardial infarction and stroke [13].

On the other hand, in addition to the aforementioned actions, drugs can act in other places and systems, also causing adverse and deleterious events in the organism. These include dry cough, taste alteration, dizziness, hypersensitivity reactions with rash (rash), angioneurotic edema, hyperkalemia, reduction of glomerular filtration with increased serum levels of urea and creatinine, and fetal complications [11]. Thus, countless events are reported from the use of these drugs, whether they are beneficial or malevolent.

2. ACEI and ARB in the management of hypertension

Assuming that the SRAA is a fundamental regulator of cardiovascular and renal functions, ACEI and ARB have different outcomes in the regulatory systems of the human organism, especially in the cardiovascular system. In recent decades, several clinical studies have confirmed that SRAA suppression reduces cardiovascular mortality and total mortality of patients [14].

Despite this, the protective role of SRAA inhibitors in relation to cardiovascular mortality was questioned by studies such as non-insulin-dependent diabetes, hypertension, microalbuminuria or proteinuria, cardiovascular events, and ramipril study (DIABHYCAR) [15] which showed that the use of ACE inhibitors had no effect on cardiovascular mortality in patients with type 2 diabetes mellitus type 2 (DM2) and albuminuria. Also, studies such as Randomized Olmesartan and Diabetes

Microalbuminuria Prevention (ROADMAP) study [16] ended up demonstrating a high rate of cardiovascular mortality in patients using Olmesartan and DM2.

In this chapter, we seek to synthesize the main prospective, randomized and double-blind clinical trials that report the outcomes resulting from the use of renin-angiotensin system inhibitors in hypertensive patients, and to extend the discussion previously made in an article published by the author [17].

The following outcomes will be analyzed: total mortality (including cardiovascular and other causes of death in this specific population), cardiovascular mortality, acute myocardial infarction, stroke, heart failure, and hospitalization due to heart failure, and an addition of all these outcomes (total outcomes).

3. ACEI × ARB: the state of the art in terms of medical evidence

The use of ACEI or ARB has always been part of the first line of treatment of patients with systemic arterial hypertension, especially in the newly diagnosed or long-term patients. Not uncommon are the revaluation of each medication, due to the inefficacy in reducing pressure values or in the presence of side effects (coughing is a very common effect of ACE inhibitors, leading to the main cause of treatment change) [9].

Thus, even with the notion that the ACEI and the ARB are equally effective in reducing pressure in patients with systemic arterial hypertension, their efficiencies compared and their advantages, disadvantages, and outcomes, related to recent or chronic use, remain unknown or liable to be questioned [11].

Related to the last major clinical trials, such as SOLVD (enalapril in heart failure), HOPE (ramipril in patients with high CVD risk), and EUROPA (perindopril in stable coronary disease) studies, we noticed that the inhibition of the SRAA in specific populations caused a significant reduction in outcomes such as general mortality, mortality from cardiovascular causes, acute myocardial infarction, and STROKE [14].

Even so, the large number of studies performed, comparing the use of ACEI or ARB in reducing outcomes in their patients, was not organized and correlated in a review format with well-established characteristics. Thus, it is perceived that the latest systematic reviews with meta-analyses and clinical trials, related to the topic, can be framed in the following situations:

- Systematic reviews, with meta-analysis, based on clinical trials that studied the outcomes related to the use of ACE inhibitors or ARB without their study population necessarily composed of essential hypertensive patients [15, 16].

- Research that studied the use of ACEI in hypertensive patients secondary to other pathologies, although the focus of the study was the evaluation of the symptomatic treatment of the disease and not the hypertension itself [17].

- Studies that used as a control group another drug, which is not considered as the first line of treatment [18].

- Studies that analyzed hypertensive patients with control groups or with standard first-line treatment, according to the hypertension guidelines used by the research team [19, 20].

This presents itself as one of the most current and systematic revisions of the area. By analyzing 20 clinical trials, involving a number of 158,998 patients, the authors were able to correlate the outcomes related to the analyzed studies, such as cardiovascular and total mortality. However, the present study did not stratify the influence of a given variable (e.g., which drug has a higher mortality rate per stroke of heart failure?), as well as its selected studies involved various types of control groups when compared to the use of ACEI or ARB [14]. Thus, there was an increase, in the last instance, of the individual heterogeneity of each patient according to their medication and particular health condition.

Another meta-analysis of Ho et al. [19] encompassed 10 clinical studies that used ACEI or ARB with its outcomes stratified in cardiovascular mortality, myocardial infarction, stroke, and overall mortality. However, it involved studies in which the population was not hypertensive in some cases, such as the Comparison of Amlodipine vs. Enalapril to Limit Occurrences of Thrombosis (CAMELOT) study, which did not present its series of outcomes due to the use of the drug but is in relation to deep vein thrombosis [21–23].

Thus, in view of these systematic reviews with meta-analyses, published in high-impact journals, it can be defined that all of them correlated the presence of outcomes, due to the use of ACEI or ARB, only in certain populations, for example, in patients with thrombosis or diabetics. Thus, only a general panorama is provided, despite the magnitude of the use of medications and their outcomes, in a population with essential hypertension. Thus, the relationship of a pathology, not necessarily related to hypertension, or the presence of a comorbid syndrome such as DM2 in hypertensive patients would present relationships with other complexes to safely affirm the direct effect of antihypertensive medication on outcomes.

However, as an attempt to increase the number of patients and studies involved in the final analysis, all used studies that presented some of the biases in the previously cited items, such as the extrapolation of hypertensive populations in groups Non-hypertensive majority, or the use of other therapies in the control group other than placebo.

4. ACEI × ARB correlation

Our analysis of the literature regarding the use of ACEI or ARB in the hypertensive population selected 17 studies over the past 10 years, covering a total of 73,761 patients [17].

Of these 17 articles, 12 studies were randomized for therapy with ARB (n = 24,697) compared to control (n = 24,722), while five compared the use of ACEI (n = 12,170) with the control group (n = 12,172) [24].

All patients in the studies were considered as prehypertensive or hypertensive according to the definition of each study, and the mean base pressure ranged from 123 to 169 mmHg. The mean age of the patients ranged from 51 to 76.9 years, with the follow-up time of the studies ranging from 2.25 to 5.5 years.

4.1 Total mortality

A total of 15 studies (n = 70,983) reported events related to total mortality. Two studies [25, 26] were statistically significant in reducing mortality compared to the control group (OR = 0.831, 95% CI, [0.730–0.945], P = 0.005) and (OR = 0.563, 95% CI, [0.341–0.931], P = 0.025), respectively. In addition, the results showed a significant total difference between ACEI and control group (OR = 0.851, 95% CI,

[0.776–0.935], P = 0.001). In relation to the ARB and control group, no significance was evidenced (OR = 1024, 95% CI, [0.960–1.091], P = 0.471) (**Figure 1**).

4.2 Cardiovascular mortality

The cardiovascular mortality was analyzed in 13 studies (n = 64,460). One study [27] showed a significant reduction in the cardiovascular mortality for ACEI compared to the control group (OR = 0.733, 95% CI, [0.625–0.860], P = 0.000).

As for the ARB, the opposite was observed in one study, with an increase in the risk in the event when comparing with the control group [28] (OR = 5.011, 95% CI, [1.449–17.334], P = 0.011).

Thus, the studies with ACEI showed a reduction in the cardiovascular mortality (OR = 0.775, 95% CI, [0.689–0.872], P = 0.000), while for the use of ARB there was no significant reduction compared to the control group (OR = 0.947, 95% CI, [0.849–1.056], P = 0.324) (**Figure 2**).

4.3 Acute myocardial infarction

Regarding the occurrence of AMI, two studies evaluating ACEI were beneficial for the reduction of the event (OR = 0.785, 95% CI, [0.689–0.895], P = 0.000), (OR = 0,769, 95% CI, [0.658–0.899], P = 0.001). Related to the therapy with ARB, only one study showed statistical significance to decrease AMI when compared to the control group (E-COST, 2005) (OR = 0.428, 95% CI, [0.202–0.905], P = 0,026). A total of 14 studies (n = 67,237) showed that both ACEI and ARB are positive in the reduction of such event when compared to the control group therapy, with greater significance for ACEI (OR = 0.784, 95% CI, [0.712–0.863], P = 0.000) on ARB (OR = 0.906, 95% CI, [0.834–0.985], P = 0.020) (**Figure 3**).

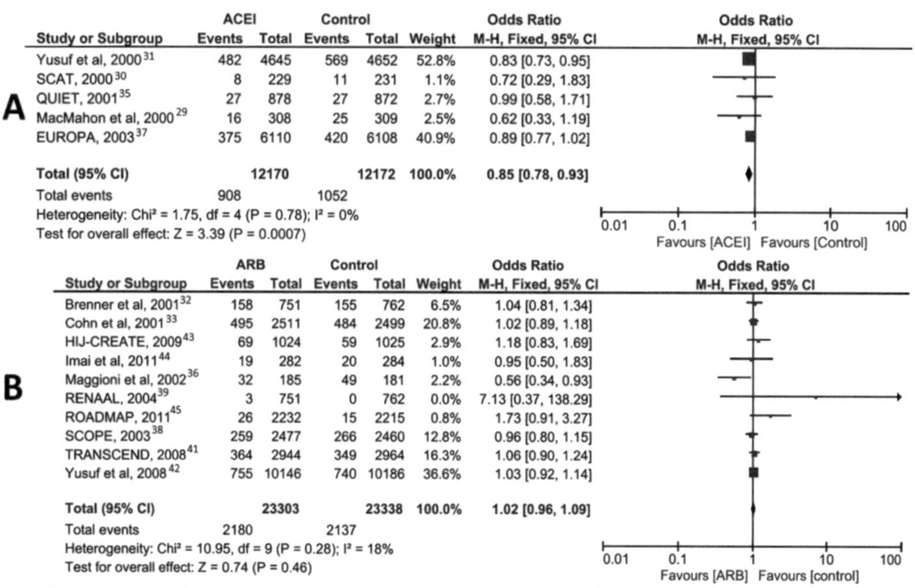

Meta-analysis using the size of the population of the studies for A: Use of ACEI and B: Use of ARB. SCAT = Simvastatin/Enalapril Coronary Atherosclerosis Trial, EUROPA = European Trial On Reduction of cardiac events with Perindopril in stable coronary Artery disease, RENAAL = Randomized Evaluation of Normal versus Augmented Level Replacement Therapy, ROADMAP = Randomized Olmesartan and Diabetes Microalbuminuria Prevention, SCOPE = The Study on Cognition and Prognosis in the Elderly, TRANSCEND = Telmisartan Randomized Assessment Study in ACE-Intolerant Subjects with Cardiovascular Disease.

Figure 1.
Forests plots summarizing the correlation of (A) ACEI and (B) ARB use for overall mortality.

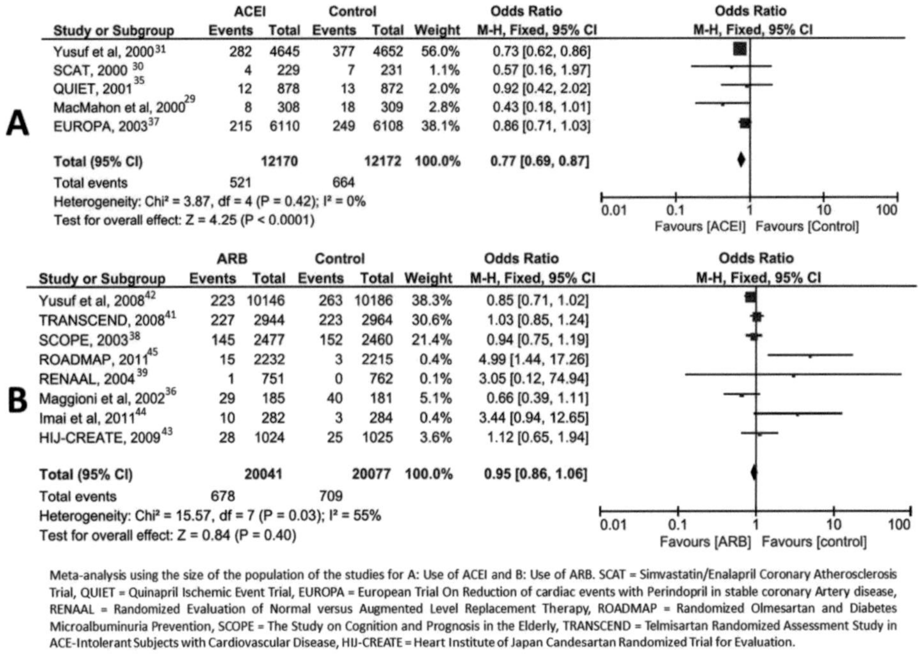

Figure 2.
Forest plots summarizing the correlation of (A) ACEI and (B) ARB use for cardiovascular mortality.

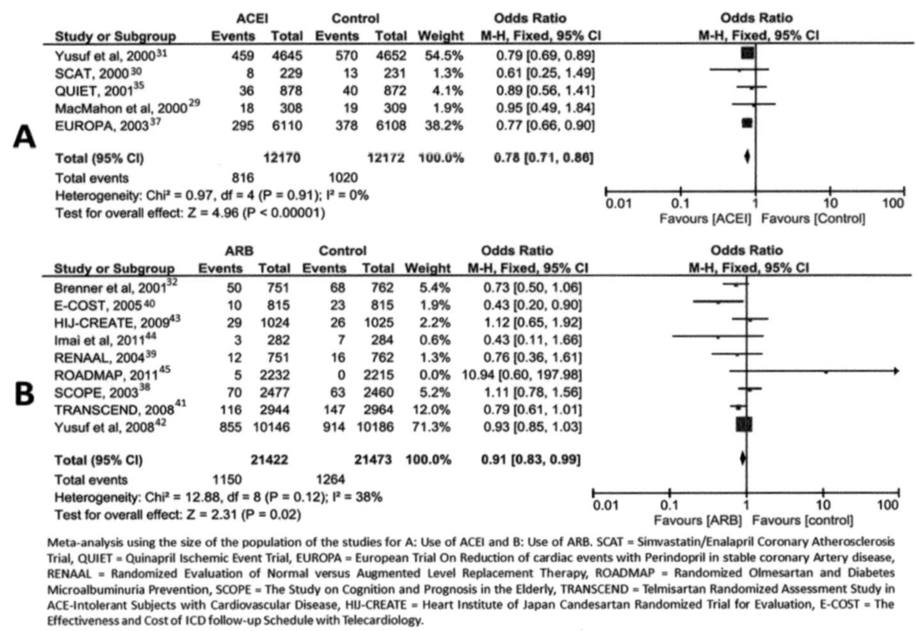

Figure 3.
Forest plots summarizing the correlation of (A) ACEI and (B) ARB use for acute myocardial infarction.

4.4 Stroke

Of the 13 studies (n = 65,724) that evaluated stroke, 2 of them, 1 involving ACEI (OR = 0.681, 95% CI, [0.553–0.838], P = 0.000) and another involving ARB (OR = 0.587, 95% CI, [0.402–0.855], P = 0.005), showed statistical significance for reduction of stroke when compared to the control group. Similar to the results for AMI, the reduction of stroke, compared to the control group, was significative in the two classes of medications, with a more expressive reduction for ACE (OR = 0.770, 95% CI, [0.654–0.908], P = 0.002) on BRA (OR = 0,890, 95% CI, [0.820–0.965], P = 0.005) (**Figure 4**).

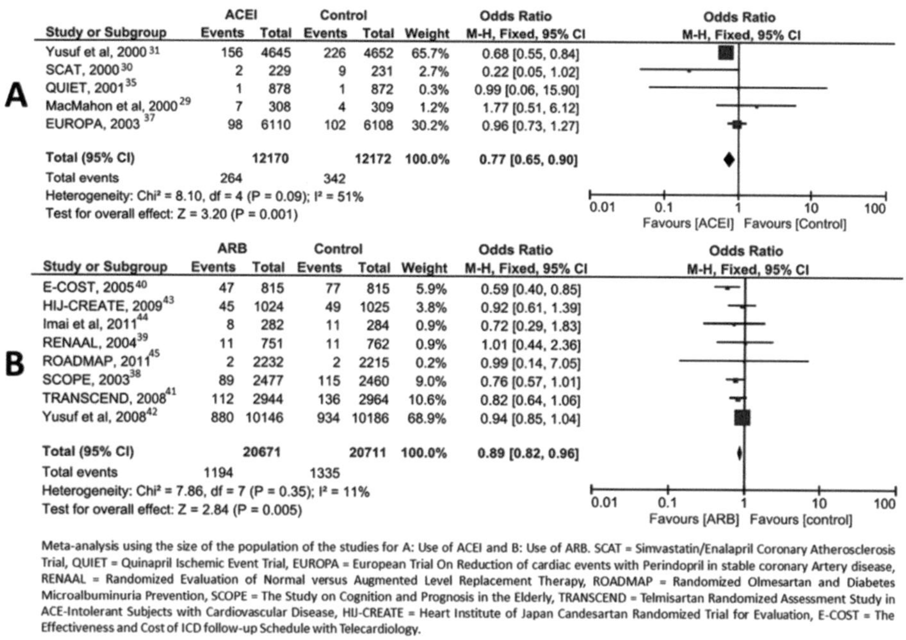

Figure 4.
Forests plots summarizing the correlation of (A) ACEI and (B) ARB use for stroke.

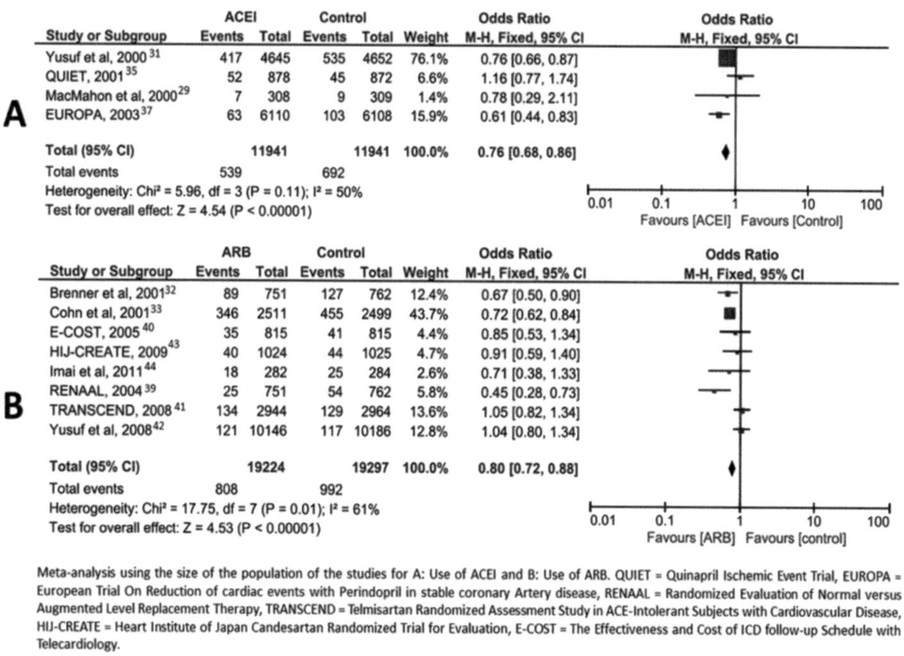

Figure 5.
Forests plots summarizing the correlation of (A) ACEI and (B) ARB use for heart failure.

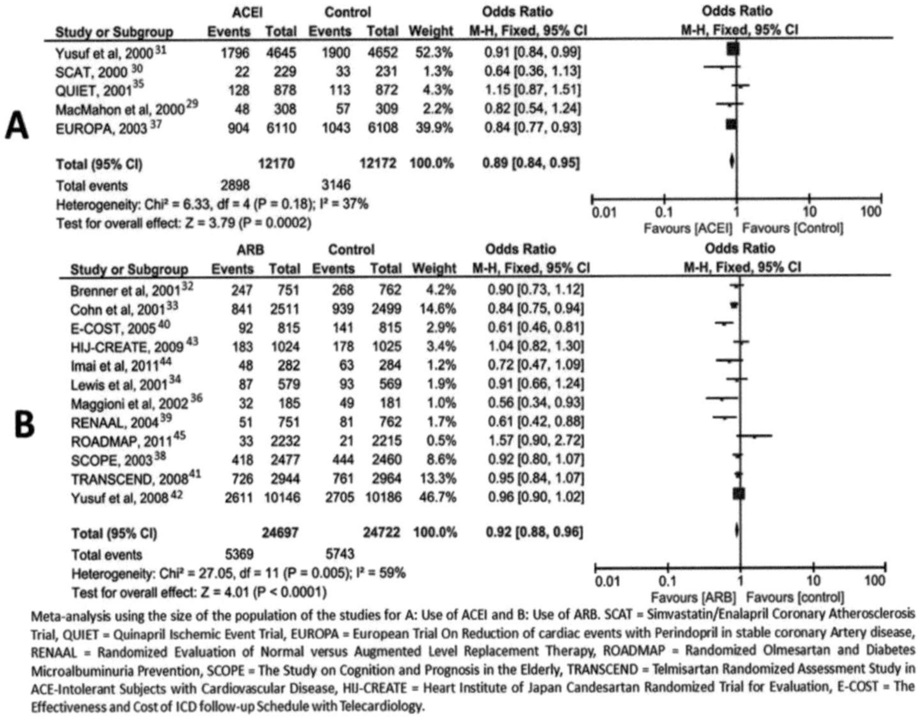

Figure 6.
Forests plots summarizing the correlation of (A) ACEI and (B) ARB use for all the outcomes.

4.5 Heart failure

A total of 12 studies (n = 62,403) analyzed heart failure. Two comparing the use of ACEI with a control group caused a significant reduction of the outcome, (OR = 0.759, 95% CI, [0.663–0.869], P = 0.000] and (OR = 0.607, 95% CI, [0.443–0.833], P = 0.002). The same was observed for three studies with ARB (OR = 0.718, 95% CI, [0.617–0.836], P = 0.000), (OR = 0.672, 95% CI, [0.502–0.900], P = 0.008) and (OR = 0.451, 95% CI [0.278–0.734], P = 0.001). Thus, the two therapies analyzed were significant in the reduction of the event, with proximity of both classes in relation to the reduction of heart failure compared to the control group: ACEI (OR = 0,762, 95% CI, [0.677–0.857], P = 0.000) and BRA (OR = 0,799, 95% CI, [0.724–0.882], P = 0.000) (**Figure 5**).

4.6 Total outcomes

Analyzing all the outcomes described, the 17 studies (n = 73,761) showed that both the uses of ACEI (OR = 0.890, 95% CI, [0.837–0.945], P = 0.000) and of ARB (OR = 0.916, 95% CI, [0.877–0.956], P = 0.000) are favorable to a significant reduction in the occurrence of mortality events (**Figure 6**).

5. Discussion

The blockade of the renin-angiotensin-aldosterone system, regardless of the level in which it is performed, by itself, is a demonstrably relevant factor in lowering blood pressure. Both the inhibition of angiotensin-converting enzyme (ACEI) and the blockade of the AT-1 receptor of angiotensin II (ARB) provide a decrease in the smooth muscle contraction in the arterial endothelium and thus lead to a reduction in the pressure values in both drug regimens [5].

With this basis, associated with the data presented in the results, it was already expected that, when comparing the use of ACEI or ARB with the control group, there was a decrease in the development of myocardial infarction, stroke, and heart failure (in favor of the first group [ACEI or ARB]). The decrease in blood pressure constitutes an agent that directly causes an improvement in the circulatory system and in the perfusion system, and, as a consequence, there is a decrease in vascular pathologies, especially in the present study, for these outcomes. Thus, the first plausible analysis of conclusion is that both the uses of ACEI and the ARB are efficient in decreasing blood pressure values as well as in the decrease of the hard outcomes. Quantitatively, it is noteworthy that the values, in these three outcomes, are more relevant in favor of the use of ACEI, when compared with the use of ARB, as seen in the charts of **Figures 4–6**. Despite this numerical superiority, in medical practice, both medications are relevant and preventive in terms of the three outcomes and therefore have their use indicated.

However, despite the synergistic effects of these medications in the hypertensive condition, their association is contraindicated. This contraindication is based on the fact that its association, when compared with its isolated use, does not cause improvement of cardiovascular outcomes, although there is an additional reduction in blood pressure [25]. In addition, the concomitant use of both medications presents, based on concrete evidence, an increase of at least 10% in the relative risk of cancer [26].

However, when analyzing the total mortality and cardiovascular mortality, there were differences. For these outcomes, as shown in **Figures 2** and **3**, the use of ACEI proved to be significant in relation to the statistic, while the use of ARB showed no differences compared to the control group. This fact, regarding the total mortality, corroborates with the data presented by van Vark [16] where there was a 10% decrease in mortality due to all causes, over 4 years, with the use of ACEI, while the use of ARB does not rush any impact. However, related to the cardiovascular mortality, this same review did not present differences between the two drugs, whereas, in our review, the differences were statistically significant [14].

Given this, a reduction in the values of overall and cardiovascular mortalities with the use of ACEI, and in view of the follow-up time of the studies allocated here (from 27 to 66 months), we relate this difference observed to an increase or decrease in the serum concentration of some endogenous molecular factor, i.e., produced by the angiotensin-renin system itself.

Based on the physiological analysis of the blockade of ACEI and ARB drugs in the angiotensin-aldosterone cascade, we found that the role of ACE inhibitors occurs by inhibiting the conversion of angiotensin I into angiotensin II, while the ARB acts later in this process, blocking the receptor of the molecule of angiotensin II. Thus, there is a decrease in the serum values of angiotensin II in the use of ACEI, which does not occur with the use of ARB [27].

Based on this mechanism of action, the use of ACEI ultimately provides higher levels of polypeptide hormones, being the most physiological importance in this context, the bradykinin [28].

Bradykinin is an inflammatory polypeptide, which is responsible for the maintenance of the inflammation cascade as well as for providing vasodilation, a necessary step so that the molecular and cellular agents of the inflammatory cascade can act on the site of possible injury. Its inactivation is performed mainly by carboxypeptidase II (kininase type II) of the lung parenchyma. Coincidentally, the kininase II, which inactivates bradykinin, is the same inhibited by the ACEI. Therefore, these medications, in addition to decreasing the concentration of angiotensin II, a potent vasoconstrictor, collaborate to increase the pool of bradykinin, a physiological inflammatory vasodilator. Thus, there is an even greater increase in the hypotensive effects as well as a greater synergism regarding the reduction of other outcomes from vasoconstriction, caused by hypertensive conditions [29–31].

Thus, the benefits of using ACE inhibitors, resulting from the increase in the amount of bradykinin in the body, seem to have an additional benefit in the population with essential hypertension. This fact stems from the knowledge, about 80 years ago, of the modulating roles of bradykinins in the renin-angiotensin system, not only in the mechanism of vasodilation but also in the ventricular remodeling in patients with hypertension [32, 33].

As for the other outcomes such as infarction, stroke, and heart failure, the increase in circulating bradykinin in these patients, due to its already explained vasodilating properties, also has a positive effect. Among these effects, we highlight the performance in the maintenance of the coronary reserve as well as in the consequent prevention of infarction [34]. In addition, bradykinin was also identified as an endogenous factor that, associated with other molecules, favors the maintenance of myocardial perfusion and consequently promotes the prevention of ventricular remodeling in heart failure [34].

Related to the prevention of stroke, as was verified in the quantitative difference between these outcomes when compared to the population using ACEI, the increase in bradykinin implies the increase of the cerebral arterial flow [35, 36].

However, although we observe that the use of ACEI is intimately related to the decrease in infarction, stroke, and heart failure, these effects do not present a significant difference when compared to the use of ARB (**Figures 3–5**).

This characteristic is due to the fact that, even if the ACEIs have a significant increase in the concentration of bradykinins, its use is also implicated in the increase of other molecular factors which, in turn, present a myriad of receptors with often antagonistic functions. This may facilitate, hinder, or even have no effect on the action of the bradykinin in those outcomes but rather present more complex interactions in the function of these pathologies [37–39].

Some of these substances have already been discovered, such as prostaglandins and prostacyclins. However, the cellular molecule-receptor interaction and the best detailing of its cardiovascular-protective properties still lack studies to discover its relations with bradykinin and, above all, the relationship between the use of ACEI and the hypertensive condition in infarction, stroke, and heart failure. Thus, so far, it is only possible to infer the main role of bradykinin and a possible nonspecific interference of these factors in such pathologies [40–42].

Now, related to total and cardiovascular mortalities, we believe that the reduction of these outcomes observed in the use of ACEI may be closely related to the increase in the bloodstream of bradykinin. This fact is justified in view of the benefits that bradykinin promotes, in the long term, in the maintenance of vascular integrity. This molecule has a positive impact on the prevention of microvascular alterations, which occur in certain diseases, in this case hypertension. This prevention tends to take place because bradykinin, via B2 receptors, protects the viability of the endothelium exposed to necrosis and apoptotic in hostile environments [43]. These situations are not observed when using the ARB. This is because its mechanism of action does not involve a pathway that changes the concentration of other mediators of blood pressure regulation, in addition to angiotensin II. Therefore, the use of ACEI is more effective than the use of ARB in these long-term outcomes.

As a consequence of this difference in the outcomes, total mortality and cardiovascular mortality, a decrease in the total outcomes with the use of ACEI was also predicted, as shown in **Figure 6**.

Moreover, analyzing the consumption of ACEI and ARB, it is observed that in recent years the consumption of ACEI decreased, while the ARB showed significant growth. In the face of the literature and the present study, it can be corroborating that this situation does not present a relevant scientific basis [44].

6. Limitations of the ACEI × ARB analysis

The main limitation of this review involves the variation between the population samples, such as the different doses of the medications and the control group, and the studies with different follow-up times. In addition, some studies called therapy as a placebo without exactly specifying this information.

Also, for medical and ethical reasons, the vast majority of studies made it clear that both groups in use of ACEI and in use of placebo were already using other antihypertensive drugs. Thus, it cannot be inferred whether these medications in use could influence the information presented here.

It is also noteworthy that the study was based on analyzing the outcomes of different classes, ACEI and ARB. Thus, it should be considered that differences between the various drugs of each class are possible.

7. Conclusion

Through the use of ACEI and ARB, the pressure levels present significant reductions with both medications. Also, in the long term, the two therapies provide a decrease in the risk of hard outcomes such as acute myocardial infarction, stroke, and heart failure. However, the use of ACEI is effective in the reduction of the overall and cardiovascular-related mortality, facts not seen using the ARB. Thus, it is assumed that this reduction is due to the higher concentration of bradykinin, in the use of ACEI, due to its cardiovascular-protective properties, and not linked to the pressure drop only.

Conflict of interest

The authors declare no conflict of interest.

Author details

Gabriel Lucca de Oliveira Salvador
Internal Medicine Department, Hospital de Clinicas, Federal University of Parana, Curitiba, Parana, Brazil

*Address all correspondence to: glucca11@gmail.com

References

[1] Kornelia K, Dirk B, Guy B, et al. Lifestyle and risk factor management in people at high risk of cardiovascular disease. A report from the European Society of Cardiology European Action on Secondary and Primary Prevention by Intervention to Reduce Events (EUROASPIRE) IV cross-sectional survey in 14 European regions. European Journal of Preventive Cardiology. 2016;**23**:1-96

[2] Wang W, Lee ET, Fabsitz RR, et al. A longitudinal study of hypertension risk factors and their relation to cardiovascular disease: The Strong Heart Study. Hypertension. 2006;**47**:403-409

[3] Montalescot G et al. 2013 ESC guidelines on the management of stable coronary artery disease: The task force on the management of stable coronary artery disease of the European Society of Cardiology. European Heart Journal. 2013;**34**:2949-3003

[4] Padwal R. Evidence based management of hypertension: Cardiovascular risk factors and their effects on the decision to treat hypertension: Evidence based review. BMJ. 2001;**322**:977-980

[5] Evora PRB, Nather JC, Rodrigues AJ. Prevalence of heart disease demonstrated in 60 years of the Arquivos Brasileiros de Cardiologia. Arquivos Brasileiros de Cardiologia. 2014;**102**:3-9

[6] Word Health Organization. Cardiovascular diseases (CVDs) Fact Sheet. WHO. 2012. Available from: http://www.who.int/mediacentre/factsheets/fs317/en/ [Accessed: 21 January 2015]

[7] Santulli G. Epidemiology of cardiovascular disease in the 21st century: Updated numbers and updated facts. American Journal of Cardiovascular Disease. 2013;**1**:1-2

[8] Chobanian AV, Bakris GL, Black HR, et al. Seventh report of the Joint National Committee on prevention, detection, evaluation, and treatment of high blood pressure. Hypertension. 2003;**42**:1206-1252

[9] Katakam R, Brukamp K, Townsend RR. What is the proper workup of a patient with hypertension? Cleveland Clinic Journal of Medicine. 2008;**75**:663-672

[10] Institute for Clinical Systems Improvement (ICSI). Hypertension Diagnosis and Treatment. Bloomington, Minn: Institute for Clinical Systems Improvement (ICSI); 2010. Available from: https://www.icsi.org/_asset/wjqy4g/HTN.pdf [Accessed: 01 June 2017]

[11] James PA, Oparil S, Carter BL, et al. 2014 evidence-based guideline for the management of high blood pressure in adults: Report from the panel members appointed to the Eighth Joint National Committee (JNC 8). JAMA. 2014;**311**:507-520

[12] Wood S. JNC 8 at last! Guidelines ease up on BP thresholds, drug choices. Heartwire. 2013. Available from: http://www.medscape.com/viewarticle/817991 [Accessed: 18 December 2013]

[13] Hajjar I, Kotchen TA. Trends in prevalence, awareness, treatment, and control of hypertension in the United States, 1988-2000. JAMA. 2003;**290**:199-206

[14] Sarah-Jo S, Laurie AT, Adrian AR, et al. Comparative effectiveness of fourth-line anti-hypertensive agents in resistant hypertension: A systematic review and meta-analysis. European

Journal of Preventive Cardiology. 2016;**24**:228-238

[15] Kornelia K, Dirk B, Guy B, et al. Lifestyle and risk factor management in people at high risk of cardiovascular disease. A report from the European Society of Cardiology European Action on Secondary and Primary Prevention by Intervention to Reduce Events (EUROASPIRE) IV cross-sectional survey in 14 European regions. European Journal of Preventive Cardiology. 2016;**23**:2007-2018

[16] van Vark LC, Bertrand M, Akkerhuis KM, et al. Angiotensin-converting enzyme inhibitors reduce mortality in hypertension: A meta-analysis of randomized clinical trials of renin-angiotensin-aldosterone system inhibitors involving 158,998 patients. European Heart Journal. 2012;**33**:2088-2097

[17] Baker WL, Coleman CI, Kluger J, et al. Systematic review: Comparative effectiveness of angiotensin-converting enzyme inhibitors or angiotensin II-receptor blockers for ischemic heart disease. Annals of Internal Medicine. 2009;**151**:861-871

[18] Hao G, Wang Z, Guo R, et al. Effects of ACEI/ARB in hypertensive patients with type 2 diabetes mellitus: A meta-analysis of randomized controlled studies. BMC Cardiovascular Disorders. 2014;**14**:148

[19] Ong HT, Ong LM, Ho JJ. Angiotensin-converting enzyme inhibitors (ACEIs) and angiotensin-receptor blockers (ARBs) in patients at high risk of cardiovascular events: A meta-analysis of 10 randomised placebo-controlled trials. ISRN Cardiology. 2013;**151**:861-871

[20] Ferrari R, Boersma E. The impact of ACE inhibition on all-cause and cardiovascular mortality in contemporary hypertension trials: A review. Expert Review of Cardiovascular Therapy. 2013;**11**:705-717

[21] JPT H, Green S, editors. Cochrane Handbook for Systematic Reviews of Interventions Version 5.1.0. The Cochrane Collaboration; 2011 Available from: www.cochrane-handbook.org

[22] Moher D, Liberati A, Tetzlaff JAD, PRISMA group. Preferred reporting items for systematic reviews and meta analyses: The Prisma statement. PLoS Medicine. 2009;**6**:1-15

[23] Mancia G, Fagard R, Narkiewicz K, et al. European society of Hypertension/European society of cardiology guidelines for the management of arterial hypertension. European Heart Journal. 2013;**34**:2159-2219

[24] Robinson KA, Dickersin K. Development of a highly sensitive search strategy for the retrieval of reports of controlled trials using PubMed. International Journal of Epidemiology. 2002;**31**:150-153

[25] ClinicalTrials.gov. Why should i register and submit results? 2011. Available from: https://www.clinicaltrials.gov/ct2/manage-recs/background [Accessed: 25 August 2015]

[26] U.S. Food and Drug Administration. Food and Drug Administration Amendments Act (FDAAA) of 2007. 2007. Available from: https://www.gpo.gov/fdsys/pkg/PLAW-110publ85/pdf/PLAW-110publ85.pdf [Accessed: 30 May 2016]

[27] International Clinical Trials Registration Platform. International Standards for Clinical Trials Registries. 2012. Available from: http://apps.who.int/iris/bitstream/10665/76705/1/9789241504294_eng.pdf [Accessed: 30 May 2016]

[28] Higgins JPTAD, Sterne JAC. Assessing risk of bias in included

studies. In: Higgins JPT, Green S, eds. Cochrane Handbook for Systematic Reviews of Interventions. Version 5.1.0. Cochrane Collaboration, 2011. Available from: http://training.cochrane.org/handbook [Accessed: 02 May 2015]

[29] Hedges LV, Vevea JL. Fixed- and random-effects models in meta-analysis. Psychological Methods. 1998;**3**:486-504

[30] MacMahon S, Sharpe N, Gamble G, et al. Randomized, placebo-controlled trial of the angiotensin-converting enzyme inhibitor, ramipril, in patients with coronary or other occlusive arterial disease. Journal of the American College of Cardiology. 2000;**36**:438-443

[31] Teo KK, Buller CE, Plante S, et al. Long-term effects of cholesterol lowering and angiotensin-converting enzyme inhibition on coronary atherosclerosis: The Simvastatin/Enalapril Coronary Atherosclerosis Trial (SCAT). Circulation. 2000;**102**:1748

[32] Yusuf S, Sleight P, Pogue J, et al. Effects of an angiotensin-converting-enzyme inhibitor, ramipril, on cardiovascular events in high-risk patients. The New England Journal of Medicine. 2000;**342**:145-153

[33] Brenner BM, Cooper ME, de Zeeuw D, et al. Effects of losartan on renal and cardiovascular outcomes in patients with type 2 diabetes and nephropathy. The New England Journal of Medicine. 2001;**345**:861-869

[34] Cohn JN, Tognoni G. A randomized trial of the angiotensin-receptor blocker valsartan in chronic heart failure. The New England Journal of Medicine. 2001;**345**:1667-1675

[35] Lewis EJ, Hunsicker LG, Clarke WR, et al. Renoprotective effect of the angiotensin-receptor antagonist Irbesartan in patients with nephropathy due to type 2 diabetes. The New England Journal of Medicine. 2001;**345**:851-860

[36] Pitt B, O'Neill B, Feldman R, et al. The quinapril ischemic event trial (QUIET): Evaluation of chronic ACE inhibitor therapy in patients with ischemic heart disease and preserved left ventricular function. The American Journal of Cardiology. 2001;**87**:1058-1063

[37] Maggioni AP, Anand I, Gottlieb SO, et al. Effects of valsartan on morbidity and mortality in patients with heart failure not receiving angiotensin-converting enzyme inhibitors. Journal of the American College of Cardiology. 2002;**40**:1414-1421

[38] Fox KM. Efficacy of perindopril in reduction of cardiovascular events among patients with stable coronary artery disease: Randomised, double-blind, placebo-controlled, multicentre trial (the EUROPA study). Lancet. 2003;**362**:782-788

[39] Lithell H, Hansson L, Skoog I, et al. The study on cognition and prognosis in the elderly (SCOPE): Principal results of a randomized double-blind intervention trial. Journal of Hypertension. 2003;**21**:875-886

[40] Remuzzi G. Continuum of Renoprotection with losartan at all stages of type 2 diabetic nephropathy: A post hoc analysis of the RENAAL trial results. Journal of the American Society of Nephrology. 2004;**15**:3117-3125

[41] Suzuki H, Kanno Y. Effects of candesartan on cardiovascular outcomes in Japanese hypertensive patients. (E-COST) group. Official journal of the Japanese Society of Hypertension. 2005;**28**:307-314

[42] Transcend Group. Effects of the angiotensin-receptor blocker telmisartan on cardiovascular events in high-risk patients intolerant to angiotensin-converting enzyme inhibitors: A randomised controlled trial. The Lancet. 2008;**372**:1174-1183

[43] Yusuf S, Diener H-C, Sacco RL, et al. Telmisartan to prevent recurrent stroke and cardiovascular events. The New England Journal of Medicine. 2008;**359**:1225-1237

[44] Kasanuki H, Hagiwara N, Hosoda S, et al. Angiotensin II receptor blocker-based vs. non-angiotensin II receptor blocker-based therapy in patients with angiographically documented coronary artery disease and hypertension: The Heart Institute of Japan Candesartan Randomized Trial for Evaluation in Coronary Artery Disease (HIJ-CREATE). European Heart Journal. 2009;**30**:1203-1212

Chapter 5

The Intratubular and Intracrine Renin-Angiotensin System in the Proximal Tubules of the Kidney and Its Roles in Angiotensin II-Induced Hypertension

Xiao C. Li, Ana Paula de Oliveira Leite, Xu Chen, Chunling Zhao, Xiaowen Zheng, Jianfeng Zhang and Jia L. Zhuo

Abstract

The kidney plays a fundamental role in the physiological regulation of basal blood pressure and the development of hypertension. Although the mechanisms underlying hypertension are very complex, the renin-angiotensin system (RAS) in the kidney, especially intratubular and intracellular RAS, undoubtedly plays a critical role in maintaining basal blood pressure homeostasis and the development of angiotensin II (ANG II)-dependent hypertension. In the proximal tubules, ANG II activates two G protein-coupled receptors, AT_1 and AT_2, to exert powerful effects to regulate proximal tubular sodium and fluid reabsorption by activating cell surface as well as intracellular AT_1 receptors. Increased production and actions of ANG II in the proximal tubules may cause salt and fluid retention, impair the pressure-natriuresis response, and consequently increase blood pressure in hypertension. The objectives of this chapter are to critically review and discuss our current understanding of intratubular and intracellular RAS in the kidney, and their contributions to basal blood pressure homeostasis and the development of ANG II-dependent hypertension. The new knowledge will likely help uncover novel renal mechanisms of hypertension, and develop kidney- or proximal tubule-specific strategies or drugs to prevent and treat hypertension in humans.

Keywords: angiotensin II, blood pressure, hypertension, kidney, proximal tubule

1. Introduction

According to the most recent American College of Cardiology (ACC)/ American Heart Association (AHA) reports, 46% of U.S. adults now develop hypertension and take antihypertensive drugs in their lifetime [1, 2]. Prevention and treatment of hypertension and its target organ complications cost several hundreds of billion dollars a year to the U.S. economy [3–6]. Although the causes of hypertension are multifactorial, the activation of circulating (endocrine), tissue (paracrine) and intracellular (intracrine) RAS via angiotensin II (ANG II) remains one of most important contributing mechanisms [1–7]. Indeed, angiotensin-converting enzyme (ACE) inhibitors, ANG II receptor blockers (ARBs), and renin inhibitors, which block the RAS at the enzymatic or receptor levels, are widely used to treat hypertension, reduce cardiovascular and renal disease risks, and prevent target organ damage [1–7]. However, clinical trials have shown that not all RAS-targeting drugs have the same efficacy of blocking the actions of ANG II and afford the same degree of cardiovascular, blood pressure and renal protection [1–6]. Some patients continue to develop cardiovascular and renal complications despite being treated with one or more than two of these blockers [7, 8]. The underlying mechanisms responsible for these clinical observations are not well understood. One of the possibilities may be that not all ARBs have the same ability to enter the cells to block intracellular ANG II. Some, but not all, ARB(s) such as telmisartan and losartan may exert therapeutic effects beyond the classic ARBs' properties.

There is accumulating evidence that ANG II acts not only as an endocrine or paracrine hormone activating cell surface ANG II receptors, but also as an intracellular or intracrine peptide activating intracellular ANG II receptors, though the precise roles of the latter remain largely unknown [9–11]. Indeed, in addition to activating cell surface ANG II receptors, circulating and paracrine ANG II can readily enter the cells via AT_1 receptor-mediated endocytosis. The ANG II/AT_1 receptor complex internalized into endosomes may continue to transmit signals from endosomes or be translocated to the nucleus to induce long-lasting genomic effects [12, 13]. Recently, we and others have used innovative in vitro cell expression system [14–16], in vivo adenoviral gene transfer of an intracellular ANG II protein selectively in proximal tubule cells of the rat and mouse kidneys [17, 18], or genetically modified mouse models to investigate the physiological roles and mechanisms of actions of intratubular and intracellular ANG II in the proximal tubules of the kidney, with a focus on basal blood pressure homeostasis and ANG II-induced hypertension [19, 20]. Specifically, we have determined whether intracellular ANG II is derived from AT_1 (AT_{1a}) receptor-mediated uptake by the proximal tubule cells, and whether proximal tubule-selective expression of an intracellular ANG II fusion protein in the rat and mouse kidney increases the expression and activity of NHE3, promotes proximal tubular sodium and fluid reabsorption, and therefore elevates arterial blood pressure [17–23]. These new studies have generated new knowledge to improve, and provided new insights into our understanding of renal mechanisms of hypertension involving both endocrine, paracrine and intracellular ANG II, and perhaps aid the development of new classes of multifunctional drugs to treat ANG II-induced hypertension and its target organ damage by blocking not only extracellular but also intracellular and nuclear actions of ANG II. Accordingly, the objectives of this chapter are to critically review, analyze, and discuss the recent developments and progresses in the studies of novel renal mechanisms of hypertension with a focus on the roles of intratubular and intracellular ANG II in the proximal tubules of the kidney.

2. Localization of intratubular and intracellular RAS and its receptors in the proximal tubules of the kidney

2.1 Angiotensinogen

Angiotensinogen, a ~60 kDa α2 globulin in the serpin family, is the primary, if not the only, substrate for the RAS super family. It is well-recognized that angiotensinogen is primarily expressed or produced in the liver under physiological conditions. Human angiotensinogen consists of 452 amino acids, whereas rodent's angiotensinogen may vary in its molecular size slightly from human form [24–27]. Angiotensinogen, not active in itself, is released from the liver and cleaved in the circulation by the rate-limiting enzyme renin to form the still inactive decapeptide ANG I. This is followed by the conversion of inactive ANG I to the active and potent peptide ANG II, initiating important biological and physiological actions. A second enzyme called angiotensin I-converting enzyme (ACE) acts to convert ANG I to form the biologically active ANG II, initiating an important biochemical and physiological angiotensinogen/renin/ANG I/ACE/ANG II cascade (see below section on ACE). Accordingly, the recognized and primary role of angiotensinogen is to serve as a key substrate to the production of ANG II in the circulation and tissues.

In the kidney, angiotensinogen mRNAs and proteins have been localized in the kidney, primarily in the proximal tubules [28–30]. Immunohistochemistry, immunoelectron microscopy and non-isotopic hybridization histochemistry have demonstrated the localization of angiotensinogen mRNAs and proteins in the proximal convoluted and straight tubules of the cortex, with glomerular mesangial cells and medullary vascular bundles also being immunopositive in neonatal rat kidney [29, 30]. In the adult rat kidney, however, angiotensinogen mRNA expression was localized primarily in the proximal convoluted tubules, whereas electron-microscopic immunohistochemistry localized angiotensinogen immunostaining in the apical membrane of proximal convoluted tubules [29, 30]. By contrast, few if any angiotensinogen mRNAs and proteins are localized in the glomeruli, mesangial cells, or distal nephrons under physiological conditions [29, 30].

Although most of angiotensinogen in the circulation is derived from the liver, there is evidence showing that angiotensinogen is also expressed and produced in the kidney [28, 31–33]. Kobori et al. have consistently shown that angiotensinogen mRNA expression and proteins are increased in the proximal tubules of the kidney in ANG II-infused rats [28, 31–33]. However, Matsusaka et al. have demonstrated that there were no significant differences in the levels of angiotensinogen and ANG II proteins in the kidney between wildtype mice and mice with kidney-specific angiotensinogen knockout [34]. It was further found that angiotensinogen protein and ANG II levels in the kidney were nearly abolished in mice with liver-specific knockout of angiotensinogen [34]. The studies of Kobori et al. and Matsusaka et al. suggests that liver-derived angiotensinogen is the primary source of renal angiotensinogen protein and ANG II under physiological conditions, but during the ANG II-induced hypertension, angiotensinogen mRNAs and proteins are also expressed in the kidney proximal tubules.

2.2 Renin

Renin, the rate-limiting enzyme first discovered to increase blood pressure in rabbits by Tigerstedt and Bergman in 1898 [35], is an aspartyl proteinase or angiotensinogenase. Renin plays the most critical role in the initiation of the angiotensinogen/renin/ACE/ANG II/AT_1 receptor activation in the cardiovascular, kidney, and other major target tissues. Human renin precursor consists of 406 amino acids with a pre- and a pro-segment of 20 and 46 amino acids, respectively [36]. Mature human renin contains 340 amino acids and a molecular wt. of 37 kDa [36]. Renin, renin activity, and its mRNA have been localized in the kidney, submaxillary glands, blood vessels, heart, adrenal glands, and brain tissues by enzymatic assays, immunohistochemistry, in situ hybridization histochemistry etc. [37–39]. In the kidney, active renin is primarily localized in the juxtaglomerular apparatus (JGAs) in the afferent arterioles of the kidney under both physiological and diseased conditions [40–42]. For example, light and electron microscopic immunocytochemistry with an antibody to purified human renal renin localized renin in the secretion granules of the epithelioid cells of the afferent arteriole of the JGAs, in renal artery stenosis, or in Bartter's syndrome [36, 37]. In the dog kidney, we have used an in vitro autoradiographic approach to localize active renin using radiolabeled renin inhibitors [40–42]. High resolution light microscopic autoradiography specifically localized active renin to the vascular pole of the glomeruli, or the JGAs (**Figure 1**) [40–42].

In the proximal tubule of the kidney, renin mRNAs have been reported [43, 44]. Renin activity and mRNAs were detectable in cultured rabbit proximal tubule cells [45], in isolated proximal convoluted and straight tubules, but not in outer medullary collecting ducts [44]. Tang et al. reported that all major components of the RAS, including angiotensinogen, angiotensin converting enzyme, and renin, were expressed in an immortalized rat proximal tubule cell line [45]. However, there is also evidence that renin localized in the proximal tubules may be due to the uptake of circulating renin after filtration [46, 47]. Taugner et al. demonstrated that the reabsorptive pinocytosis of the filtered renin was the primary source of tubular renin in the kidney [46], whereas Iwao et al. used light and electron microscopic autoradiography to localize ^{125}I-labeled renin accumulated in the apical membranes of the proximal convoluted tubules [47]. Taken together, these studies strongly support the concept that in addition to local biosynthesis and expression, circulating or interstitial renin may be taken up by the proximal convoluted tubules in the kidney.

2.3 Angiotensin I-converting enzyme (ACE)

The 2nd key enzyme for the activation of the RAS is ACE, a dipeptidyl carboxypeptidase I, kininase II and EC 3.4.15.1 [48]. Corvol's group first molecularly cloned ACE from human vascular endothelial cells [48], whereas Bernstein's

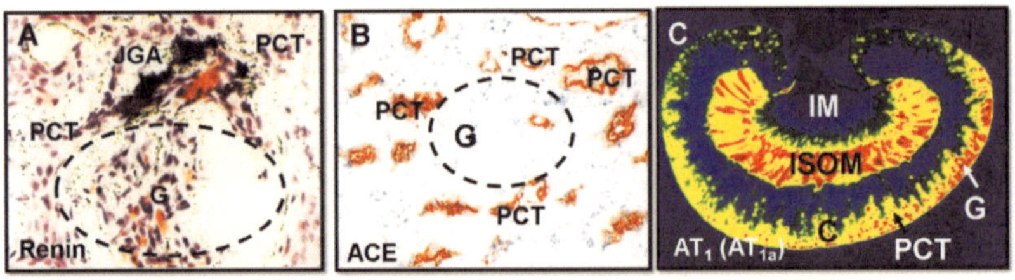

Figure 1.
Intrarenal localization of renin in the juxtaglomerular apparatus (A: JGA), angiotensin-converting enzyme (B: ACE), and angiotensin II AT_1 receptors in the kidney (C: AT_1 or AT_{1a}) using quantitative in vitro autoradiography. C, renal cortex; G, glomerulus; IM, inner medulla; ISOM, inner stripe of the outer medulla; PCT, proximal convoluted tubule.

group cloned ACE from the mouse kidney in 1988, respectively [49]. ACE in humans consists of 1306 residues with a signal peptide of 29 amino acids [48], whereas ACE in mice contains 1278 amino acids [49]. Approximately 80% of the amino acid sequences are similar between human and mouse ACE. There are two ACE isozymes, one somatic isozyme in the lung, vascular endothelial cells, renal epithelial cells, and testicular Leydig cells, and the other germinal isoenzyme solely in sperm [50–52]. The key actions of ACE are to convert the biologically inactive ANG I to the active peptide ANG II, and to degrade the vasoactive peptide bradykinin. Thus, ACE is most critical for the generation of ANG II in the circulation and tissues.

Abundant ACE is expressed and localized in the kidney, especially in the proximal tubules and glomerular and vascular endothelial cells of intrarenal blood vessels [53–57]. We and others have localized ACE proteins and its mRNA expression in the kidney using quantitative in vitro autoradiography, immunohistochemistry, and in situ hybridization histochemistry (**Figure 1**). For example, the Mendelsohn's group first localized ACE in the rat kidney using quantitative in vitro autoradiography with the radiolabeled ACE inhibitor lisinopril, ^{125}I-351A [53]. ACE was localized primarily to the inner cortex, corresponding to the proximal tubules and blood vessels [53]. We found that infusion of ANG II for 2 weeks significantly increased, rather than downregulated, ACE in the proximal tubules of the rat kidney [54]. At higher resolutions, Brunevaly et al. and others showed ACE primarily in the microvilli and brush borders of the proximal tubules in the human kidney [55–57]. In the vasculature, ACE was localized to the vascular endothelial cells especially in the peritubular capillaries, but not glomerular capillaries of the kidney [53–57]. ACE was also localized inside the renal vascular endothelial and proximal tubular cell in endoplasmic reticulum, endosomes, and nuclear envelope, suggesting the presence of intracellular and/or nuclear ACE [53–57]. However, only very low levels of ACE were detected in the inner medulla.

2.4 Angiotensin II (ANG II)

Angiotensin II (ANG II) is undoubtedly the most powerful peptide in the RAS super family, playing a key role in regulating renal blood flow, glomerular filtration, and proximal tubular reabsorption of sodium and fluid, contributing to normal blood pressure and body salt and fluid homeostasis [58–64]. It is well-recognized that the levels of ANG II in the kidney, especially in the proximal tubules, are higher than in the plasma or other tissues. Indeed, local expression and biosynthesis of angiotensinogen, renin, and ACE in the proximal tubules of the kidney significantly contribute to high levels of ANG II levels in the kidneys under physiological conditions [64–68]. Furthermore, ANG II levels are further increased in the kidney of animal models of ANG II-dependent hypertension, even though the circulating and JGA renin and ACE are suppressed [67–73]. This is likely due to the fact that the proximal tubules express all major components of the RAS necessary for the formation of ANG II [38, 47, 54, 59, 67, 74, 75], the proximal tubules have a greater capacity to take up circulating ANG II via AT_1 (AT_{1a}) receptor-mediated mechanisms [14, 19, 20, 67], and to augmentation of the expression or generation of angiotensinogen, ACE and ANG II in ANG II-induced hypertension [54, 67, 70, 73]. Finally, ANG II is not only generated in the intratubular fluid compartment, but also localized in intracellular organelles, such as endosomes, mitochondria, and nuclei [15, 67, 71, 74, 75], where it serves as an important intracellular or intracrine peptide.

2.5 AT_1 and AT_2 receptors

It is now well-accepted that ANG II binds to and activates two different classes of G protein-coupled receptors (GPCRs) to induce well-recognized cardiovascular, renal and blood pressure responses, following the successful development of nonpeptide ANG II type 1 and type 2 receptor antagonists [76–78]. Molecular cloning of AT_1 and AT_2 receptors and studies of animal models with genetically knockout of these receptors further confirms their pharmacological characterization. Murphy et al. [79] and Sasaki et al. [80] successfully cloned the AT_1 receptor in 1991, showing that the AT_1 receptor shares the seven-transmembrane-region motif of the GPCR superfamily. AT_1 receptors mediate the well-known actions of ANG II on vasoconstriction, cardiac hypertrophy, hypertensive, renal salt retention, as well as aldosterone biosynthesis [76–78, 81]. The AT_2 receptor was cloned by Mukoyama et al. [82], Nakajima et al. [83], and Kambayashi et al. [84], respectively. The AT_2 receptor was found to have 34% of the identical sequence to the AT_1 receptor, sharing a seven-transmembrane domain topology of GPCRs [82–84]. However, the roles and signal transduction pathways for the AT_2 receptor remain incompletely understood.

In the kidney, the AT_1 receptor is widely expressed and localized in different structures or cell types, most prominent in three anatomical regions, that is, the glomerulus, proximal tubules, and the inner stripe of the outer medulla, corresponding the vasa recta blood vessels and renomedullary interstitial cells (**Figure 1**) [85–87]. We and others have consistently localized the AT_1 receptor in the rodent and human kidneys using quantitative in vitro and in vivo autoradiography, with high levels of these receptors in the glomerulus, proximal tubules, and renomedullary interstitial cells (**Figure 1**) [85–87]. Other anatomical regions or renal structures may express low levels of AT_1 receptor expression, detectable with RT-PCR or immunohistochemistry. AT_1 receptors have also been localized in intracellular organelles, for example, endosomes, mitochondria, and nuclei in the proximal tubule cells, suggesting an important intracellular roles [67, 74, 88–90]. By contrast, the levels of AT_2 receptor expression in the kidney are species-related or closely associated with the kidney development. Indeed, high levels of AT_2 receptors are expressed extensively in the developing fetal and neonatal tissues, but most of them disappear before reaching the adulthood [87]. Nevertheless, the expression of AT_2 receptors appears to persist in the adrenal medulla, proximal tubules, and the adventitia of human kidney blood vessels, suggesting potential roles for these receptors in these target tissues [85–87, 91–93].

3. Intratubular and intracellular ANG II: the long-term genomic effects induced by endocrine, paracrine and intracellular ANG II

In contrast to the classic dogma that ANG II only binds to and activates cell surface GPCRs to initiate downstream signaling responses, ANG II can also bind and activate intracellular GPCRs to induce long-term genomic effects. The RAS includes an extracellular system and an intracellular system. ANG II acts as the principle effector of both extracellular and intracellular RAS. Extracellular ANG II includes circulating (endocrine) and paracrine ANG II, which plays the classical roles of the RAS through activation of cell surface GPCRs [76–78, 81, 94, 95]. Intracellular ANG II includes intracellularly formed ANG II (intracrine) and ANG II internalized through AT_1 (AT_{1a}) receptor-mediated endocytosis [96–101]. The roles of circulating and paracrine ANG II and its GPCR-mediated signaling mechanisms via cell surface receptors have been extensively investigated. By contrast, the roles of intracellular ANG II and its mechanisms of actions remain poorly understood.

This disparity in our understanding extracellular versus intracellular ANG II has led many to assume that ANG II only activates cell surface receptors to induce all of its biological and physiological responses, and that all ARBs would only block cell surface receptors to produce the same beneficial effects. Thus, an intracellular ANG II system is thought to be unnecessary in the regulation of cardiovascular, blood pressure, and renal physiology and diseases.

However, recent studies strongly suggest that these views may be revised for a number of reasons [96–101]. First, it is well-recognized that extracellular ANG II is continuously internalized with its receptors after it activates cell surface receptors. This has long been interpreted only as required for the desensitization of cell surface receptors to repetitive stimulation by extracellular ANG II by moving the ANG II/ AT_1 complex into the lysosomal pathway for degradation. There is evidence, however, that the activated agonist/receptor complex internalized into the endosomes may continue to transmit ras/mitogen-activated protein kinase (MAPK) signaling [12, 13]. Ras and MAPK signaling for AT_{1a}, vasopressin V_2, and $\beta 2$ adrenergic receptors (β2AR) have been reported in endosomal membranes [12, 13, 15, 16], the endoplasmic reticulum, the Golgi or the nucleus independent of cell surface receptor-initiated signaling [81, 88, 89, 102]. Second, ANG II exerts long-lasting genomic or transcriptional effects, which may be independent from the well-recognized effects induced by activation of cell surface receptors [97–99, 102, 103]. ANG II induces the expression or transcription of many growth factors and proliferative cytokines including nuclear factor-κB (NF-κB) [104–107], monocyte chemoattractant protein-1 (MCP-1) [106, 108], TNF-α [107], and TGF-β1 [102, 109, 110]. While hemodynamic responses to ANG II often occur in seconds or minutes, cellular growth, mitogenic, proliferative and fibrotic responses to ANG II may last from hours to weeks and months. Since the cell surface AT_1 (AT_{1a}) receptors may be desensitized in response to sustained exposure to endocrine and paracrine ANG II, the long-term genomic effects of ANG II, as observed in cardiovascular, hypertensive, and renal diseases, are at least in part mediated by intracellular ANG II system. Third, not all ARBs, ACE or renin inhibitors are created equal to block both extracellular and intracellular ANG II systems. ARBs may differ in their lipophilic ability to enter the cells to block intracellular AT_1 receptors [111–113]. Indeed, ARBs show different effects on uric acid metabolism, cell proliferation, oxidative stress, nitric oxide production and PPAR-γ activity [111–113]. We and others have shown that losartan internalized with AT_{1a} and AT_{1b} receptors, albeit at a slower rate than ANG II [19, 20, 67, 103, 114], and to attenuate ANG II-induced intracellular and nuclear effects [15, 88, 89, 102, 103]. Moreover, telmisartan not only blocks AT_1 receptors, but also acts as a partial activator of liver-specific peroxisome proliferator-activated receptor γ (PPAR-γ) [111, 115–157]. Finally, some clinical studies have shown that even treated with renin inhibitors, ACE inhibitors or ARBs, there are some patients who still progress to hypertension and suffer from cardiovascular and renal complications [111, 115–117]. These data suggest that additional mechanisms should be involved and studied accordingly. Thus, the new challenges to the field are to study whether and how intracellular ANG II may contribute to these mechanisms and design multifunctional drugs to block both extracellular and intracellular ANG II-induced effects.

4. Intratubular and intracellular ANG II: AT_{1a} receptor-mediated uptake of circulating and paracrine ANG II in the proximal tubules

We and others have investigated whether circulating and local paracrine ANG II is taken up by the proximal tubules of the kidney via AT_1 (AT_{1a}) receptor-mediated endocytosis [19, 20, 118–121], and whether internalized ANG II and AT_{1a} receptors are co-localized in the endosomal compartment and nucleus (**Figure 2**) [67, 74, 88, 89]. Our studies demonstrated that global deletion of AT_{1a} receptors blocked the uptake of unlabeled Val^5-ANG II [19] or $[^{125}I]Val^5$-ANG II in the kidney of AT_{1a}-KO mice [20]. However, these studies focused only on the entire kidney, and what nephron segments involved in taking up unlabeled Val^5-ANG II or $[^{125}I]Val^5$-ANG II could not be determined using these approaches [19, 20]. We further used cultured proximal tubules cells to test whether proximal tubule cells take up extracellular ANG II and the mechanisms involved (**Figure 2**) [14, 100, 122–126]. The advantages of using these cells for the proposed studies are that ANG II receptors are abundantly expressed and localized in both apical (AP) and basolateral (BL) membranes [127–131]. However, it has not been determined whether ANG II receptors in AP or BL membranes mediate ANG II uptake in the proximal tubules. In a previous study using a porcine proximal tubule cell line expressing a rabbit AT_1 receptor, AT_1-mediated uptake of $[^{125}I]$-ANG II was found to be significantly different between AP and BL membranes [130]. AT_1-mediated uptake of $[^{125}I]$-ANG II was more robust and efficient in AP membranes than in BL membranes [130]. Conversely, ANG II-induced AT_1 receptor internalization was reportedly much faster in BL membranes than in AP membranes of OK cells [131]. Thus these differences inAT_1-mediated uptake of $[^{125}I]$-ANG II or ANG II-induced AT_1 receptor endocytosis or internalization may underscore the differences in the cell types used or experimental conditions.

In addition to AT_1 (AT_{1a}) receptors, other factors may also regulate the uptake of extracellular ANG II by proximal tubule cells. AP membranes of proximal tubule cells express abundant endocytic receptor megalin, which plays a crucial role in mediating the uptake of low molecular weight (LMW) proteins in proximal tubule cells [132–136]. Deletion of megalin in mice led to the development of LMW proteinuria [135]. Interestingly, megalin also binds and internalizes ANG II in immortalized yolk sac cells (BN-16 cells) [136]. We have demonstrated that siRNA knockdown of megalin expression or caveolin 1 in proximal tubule cells significantly attenuated ANG II uptake by proximal tubule cells [122, 123]. However, the

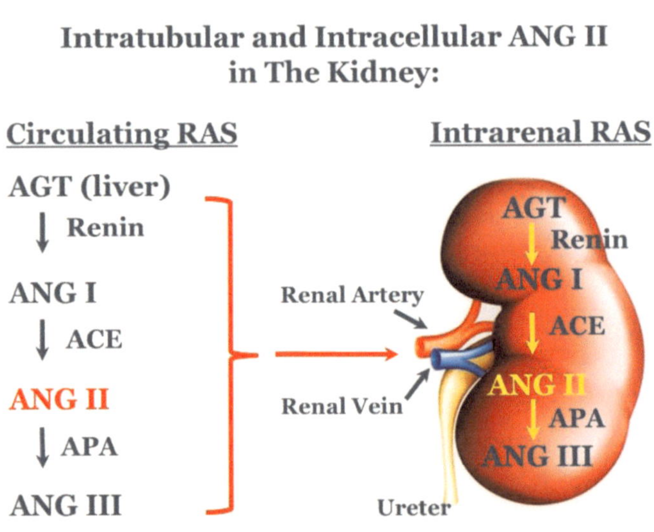

Figure 2.
All major components of the circulating RAS, including angiotensinogen (AGT), renin, angiotensin I (ANG I), and ANG II, may be filtered by the kidney glomerulus and taken up by the proximal tubules. Alternatively, all major components of the RAS may be expressed and localized in the proximal tubules of the kidney. ACE, angiotensin-converting enzyme and APA, aminopeptidase A.

extent to which megalin- and caveolin 1-mediate ANG II uptake in proximal tubule cells is significantly smaller than that mediated by AT_1 (AT_{1a}) receptor-dependent mechanism [19, 20, 122, 123].

5. Intratubular and intracellular ANG II: canonical versus noncanonical endocytic pathways in mediating ANG II uptake in the proximal tubules

We have mechanistically investigated that AT_1 (AT_{1a}) receptor-mediate the uptake of extracellular ANG II by proximal tubule cells in vitro and circulating ANG II in vivo [19, 20, 122–126]. It has been previously shown that in vascular smooth muscle cells (VSMCs), cardiomyocytes, and COS-7 cells, $\beta 2$ adrenergic receptors, AT_{1a}, epidermal growth factor receptors, and insulin receptors are internalized via the canonical clathrin-dependent pathway [137–144]. Clathrin-coated pits play an important role in invaginating and pinching off the plasma membranes to form coated vesicles and targeted to endosomes [138, 140, 142]. GPCR kinases (GRKs), small GTP-binding proteins, such as Rab5, and β-arrestins are reportedly involved in clathrin-dependent AT_{1a} endocytosis [145, 146]. However, dominant-negatives, siRNAs or knockout targeting dynamin, GRKs or β-arrestins have little effects on AT_{1a} receptor endocytosis in some studies, suggesting that alternative (non-canonical) pathways may also be involved in AT_{1a} receptor endocytosis [137–146].

There is evidence to suggest that tyrosine phosphatases may be involved in ANG II-induced AT_1 receptor endocytosis in AP and BL membranes, since the endocytic response was inhibited by the tyrosine phosphatases inhibitor, phenylarsine oxide (PAO), rather than by pertussis toxin [147–151]. Colchicine, an inhibitor of cyto-skeleton microtubules [148], also appeared to inhibit AT_1 receptor-mediated ANG II uptake and its effects in rat proximal tubule cells [150, 151]. The role of clathrin-coated pits in mediating AT_1 receptor-mediated ANG II uptake was also investigated, but we found that deletion of clathrin-coated pits with sucrose or specific siRNAs to knock down clathrin light (LC) or high chain subunits (HC) failed to alter AT_1-mediated uptake of Val^5-ANG II [151]. However, AT_1-mediated uptake of Val^5-ANG II was significantly inhibited by colchicine or siRNA knocking down of microtubule-associated proteins, MAP-1A or MAP-1B, in proximal tubule cells [151]. Our studies therefore support the scientific premise that the noncanonical microtubule-dependent endocytic pathway may be involved in mediating the AT_1-mediated uptake of ANG II in proximal tubule cells.

How ANG II and AT_1 receptors are internalized into the endosomal compartments and transported to other organelles or the nucleus in proximal tubule cells remains incompletely understood. Intravenous infusion of ^{125}I-labeled ANG II was previously detected in the nuclei of rat vascular smooth muscle cells (VSMCs) and cardiac myocytes [152] or the Golgi of adrenal cells [153]. Cook et al. showed that ANG II and its AT_{1a} receptor were translocated to the nuclei of hepatocytes and VSMCs [154]. In AT_{1a} receptor-expressing HEK 293 cells, internalized AT_{1a} receptors were detected in perinuclear areas as well as in the nuclei [155, 156]. In supporting the above-mentioned studies, we also reported high levels of internalized FITC-labeled ANG II in perinuclear areas and the nucleus, which was inhibited by colchicine and siRNA knockdown of MAP-1A [14, 122, 123, 151]. Taken together, our results strongly suggest that the microtubule-dependent pathway may play an important role in mediating the nuclear translocation of internalized ANG II/AT_1 receptor complex in proximal tubule cells. Indeed, a nuclear localization sequence (NLS, KKFKKY, aa307-312) has been identified within the AT_{1a} receptor, which may mediate nuclear trafficking and activation of AT_{1a} receptors by ANG II [155, 156].

6. Intratubular and intracellular ANG II: intracellular versus extracellular effects and signaling mechanisms in the proximal tubules

In the proximal tubules of the kidney, extracellular ANG II has been reported to stimulate the expression of Na^+/H^+ exchanger 3 (NHE3) [14, 16, 102, 125], AP insertion of NHE3 [157], Na^+/H^+ exchanger activity [158–161], or NHE3-induced $^{22}Na^+$ uptake in cultured or isolated proximal tubule cells [162, 163]. The signaling mechanisms by which extracellular ANG II increases the expression and activity of NHE3 in proximal tubule cells have been well studied and documented [164–169]. The most well-described signal mechanism is that ANG II activates cell surface receptor-coupled G proteins, with subsequent increases in IP_3 and $[Ca^{2+}]_i$, generation of DG, and activation of PKC [164–169]. The other well-recognized downstream signaling pathways for extracellular ANG II to induce biological or physiological responses also include activation or inhibition of calcium-dependent calcineurin [170], cAMP-dependent protein kinase A (PKA) [169, 171], Ca^{2+}-independent PLA_2 [172], PI 3-kinase [157], c-Src/MAP kinases ERK 1/2 [165], or nuclear factor-κB [173].

According to the principles of the G protein-coupled receptor pharmacology, ANG II must bind to its cell surface receptors to activate intracellular signaling mechanisms in order to induce responses [76–78, 138]. Upon internalization, however, ANG II may act as an intracellular peptide to induce biological or physiological responses. Indeed, blockade of the endocytosis of AT_1 receptors is associated with inhibition of PKC, IP_3 formation, and Na^+ flux in proximal tubule cells [14, 16, 122–126, 149, 150]. Furthermore, ANG II-induced AT_1 receptor endocytosis is also associated with activation of PLA_2 [147, 172], inhibition of adenylyl cyclase [151, 169, 171], and increases in Na^+ uptake from AP membranes [149–151]. We have recently shown that AT_1-mediated uptake of extracellular Val^5-ANG II was indeed associated with inhibition of basal and forskolin-stimulated cAMP accumulation [125, 151], ANG II-stimulated NHE3 expression [14, 16, 122, 123], and ANG II-induced activation of MAP Kinases ERK1/2 and nuclear factor-κB in proximal tubule cells [14, 16, 124, 126, 151].

Nevertheless, these approaches are unlikely able to distinguish the effects of ANG II mediated by cell surface or intracellular receptors. Previous studies have shown that single cell microinjection or microdialysis of ANG II directly into the cells may distinguish between the effects induced by extracellular ANG II from those induced by intracellular ANG II [15, 102, 174–177]. Indeed, we have demonstrated that intracellular microinjection of ANG II directly into single rabbit proximal tubule cells induced intracellular $[Ca^{2+}]_i$ responses (**Figure 3**) [10, 15, 16, 81, 177]. We further reported that microinjection of the AT_1 blocker losartan abolished the $[Ca^{2+}]_i$ response induced by microinjected ANG II, but it only partially blocked the effects of extracellular ANG II [15]. In further proof-of-the concept studies, we showed that ANG II stimulated nuclear AT_{1a} receptors to increase in vitro transcription of mRNAs for TGF1, MCP-1 and NHE3 in isolated rat renal cortical nuclei [102]. These studies provide evidence that intracellular ANG II may activate cytoplasmic and nuclear AT_1 receptor to induce important genomic effects in proximal tubule cells [15, 102, 174–177].

Whether intracellular ANG II may alter biological responses in a cell culture model has been determined by directly expressing an intracellular ANG II fusion

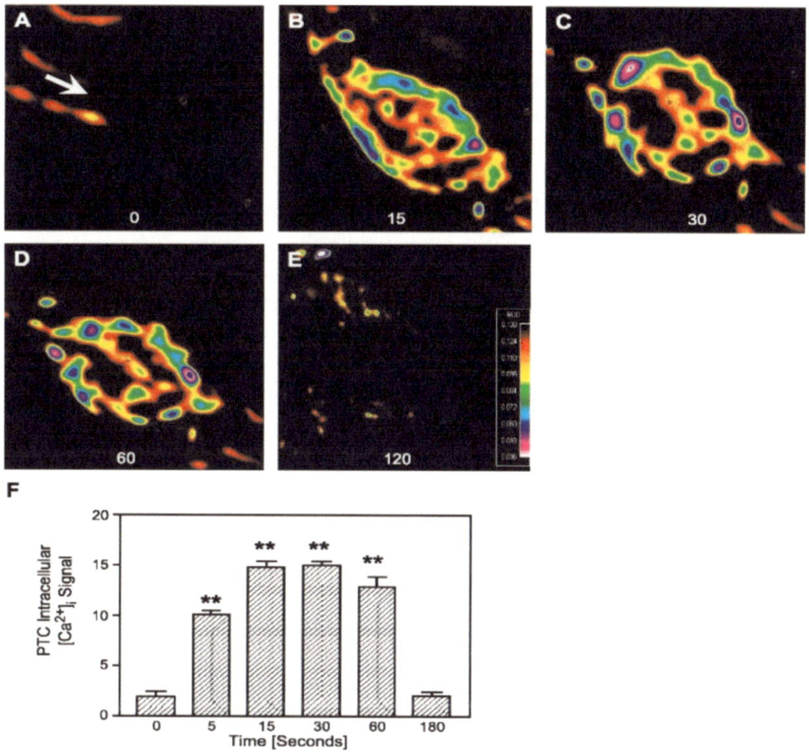

Figure 3.
Intracellular microinjection of angiotensin II induces intracellular calcium mobilization in cultured rabbit proximal tubule cells. Adapted from Zhuo et al. with permission [15].

protein [9, 11, 15, 88–90, 102]. Cook et al. overexpressed a cyan fluorescent, intracellular ANG II construct (ECFP/ANG II) with or without a rat yellow fluorescent AT_{1a} receptor (AT_1R/EYFP) in rat VSMCs or hepatocytes [9, 97, 98]. They demonstrated that intracellular ANG II induced the proliferation of VSMC via activation of cAMP response element-binding protein (CREB), p38 MAP kinase, and MAP kinases ERK 1/2 [9, 97, 98]. In another study, an intracellular ANG II (pcDNA/TO-iAng II) was expressed in CHO cells to induce cell proliferation, but none of ARBs was found to attenuate the effect of intracellular ANG II on cell proliferation [178, 179]. Nevertheless, these early proof of concept studies suggest that in vitro or in vivo expression of a cyan fluorescent intracellular ANG II fusion protein (ECFP/ANG II) in the proximal tubule cells of wild-type and AT_{1a}-KO mice may be an innovative approach to distinguish the effects of intracellular versus extracellular ANG II.

7. Intratubular and intracellular ANG II: physiological effects of intracellular versus extracellular ANG II on proximal tubule Na^+ reabsorption and blood pressure

The physiological roles of intracellular ANG II in the regulation of proximal tubule Na^+ reabsorption and normal blood pressure homeostasis remain to be determined. Whether intracellular and/or internalized ANG II may physiologically regulate proximal tubule Na^+ transport and blood pressure has not been studied until recently. Indeed, this line of research has been long stymied due to the lack of suitable animal models that express an intracellular ANG II protein, which is not secreted outside the cells and only acts intracellularly. Dr. Reudelhuber's group was the first to generate genetically modified mouse model that expresses an ANG II-producing fusion protein in the cardiomyocytes of the rat heart [180, 181]. They used the α myosin heavy chain promoter to control the expression of ANG II-releasing fusion protein in the cardiomyocytes. Cardiac specific expression of this ANG II fusion protein led to 10-fold increases in ANG II levels in the heart of these transgenic mice, but it did not elevate ANG II levels in the plasma [180, 181]. This approach is very unique to construct this cardiac-specific ANG II fusion protein with a signal peptide sequence derived from human prorenin and a furin cleavage site. Thus, the expressed ANG II fusion protein will be cleaved by furin, and released into the secretory pathway and the cardiac interstitium [180, 181]. It is expected that this cardiac-specific ANG II fusion protein activates cell surface, but not intracellular receptors. In a different study, Baker et al. expressed an intracellular ANG II peptide in the mouse cardiomyocytes using an adenoviral vector [178]. Cardiac-specific expression of this intracellular ANG II peptide in mice induced cardiac hypertrophy, but not altered blood pressure and plasma ANG II [99, 178]. Furthermore, the AT_1 receptor blocker failed to block the cardiac hypertrophic effect of this peptide, suggesting that AT_1 receptor may not be involved [99, 178].

In the kidney, a proximal tubule cell-specific promoter may be an ideal approach to express an intracellular ANG II protein selectively in the proximal tubules. For example, the kidney androgen-regulated protein gene (KAP) has been used to drive "proximal tubule-specific" expression of human angiotensinogen and renin in the kidney [182, 183]. It has been shown that the KAP gene is widely expressed in the kidney, with its expression reportedly confined to the proximal tubules and regulated by androgen and estrogen [184, 185]. The advantages of this approach are its usefulness for studying the sexual dimorphic regulation of angiotensinogen expression in the proximal tubules of the kidney [182, 183].

We have collaborated with Dr. Julie Cook of Ochsner Clinic and Dr. Isabelle Rubera of University of Nice-Sophia, France to develop an adenoviral construct

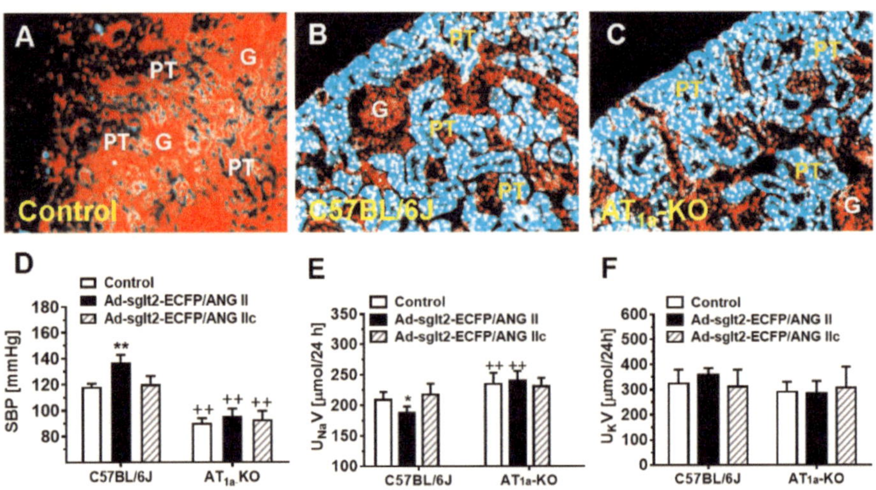

Figure 4.
Overexpression of an intracellular ECFP/ANG II fusion protein selectively in the proximal tubule of the kidney in C57BL/6J or AT_{1a}-KO mice. ECFP/ANG II increased systolic blood pressure and had a significant antinatriuretic response in C57BL/6J but not in AT_{1a}-KO mice. Green blue represents ECFP/ANG II expression in the proximal tubules, whereas Red represents DAPI-stained nuclei in the cortex after conversion from blue color. G, glomerulus. PT, proximal tubule. $^{**}p < 0.01$ versus control, whereas $^{++}p < 0.01$ versus C57BL/6J mice. Reproduced from Zhuo et al. with permission [15].

(Ad-sglt2-ECFP/ANG II), which encodes a cyan fluorescent intracellular ANG II fusion protein (ECFP/ANG II) [17, 18]. The sodium and glucose cotransporter 2 promoter, sglt2, was used to drive the expression of ECFP/ANG II selectively in the proximal tubule cells of the rat and mouse kidneys. Sglt2 is expressed almost exclusively in S1 and S2 segments of the kidney proximal tubules [186]. Using this approach, we have determined whether intrarenal adenovirus-mediated expression of intracellular ECFP/ANG II selectively in the proximal tubules of the rat and mouse kidneys increases the expression and activity of NHE3, stimulate proximal tubule sodium reabsorption, and increase blood pressure in rats and mice. We demonstrated that expression of intracellular ECFP/ANG II selectively in the proximal tubules of rats and mice significantly increased NHE3 expression, proximal tubule sodium reabsorption, and blood pressure (**Figure 4**) [17, 18]. We further showed that AT_1 receptor blocker losartan and deletion of AT_{1a} receptors in mice significantly attenuated intracellular ANG II-induced NHE3 expression, proximal tubule sodium reabsorption, and blood pressure responses, suggesting an AT_1 (AT_{1a}) receptor-mediated mechanisms.

8. Intratubular and intracellular ANG II: role of NHE3 in maintaining normal blood pressure homeostasis and ANG II-induced hypertension

The Na^+/H^+ exchanger 3 (NHE3) is the most important Na^+ transporter in AP membranes of the proximal tubules of the kidney [187–190]. NHE3 is directly and indirectly responsible for reabsorbing approximately 50–60% of filtered load of NaCl and 70–80% of filtered load of bicarbonate (HCO_3^-) [187–190]. Indeed, nearly all of the measured Na^+/H^+ exchanger activity in AP membrane vesicles of proximal tubules are mediated by NHE3 [187–190]. The importance of proximal tubule NHE3 in maintaining body salt and fluid balance and blood pressure homeostasis has not been well studied until recently. Overall, global deletion of the NHE3 gene in all tissues of mice (*Nhe3*$^{-/-}$) leads to ~50% decreases in fluid, Na^+ and HCO_3^- absorption in proximal convoluted tubules, causes salt wasting from the digestive system, and significantly decreases basal blood pressure [191–194]. One of striking phenotypes is absorptive defects in the small intestines due to intestinal NHE3 deletion [191–194]. Moreover, the transgenic rescue of the NHE3 transgene in small intestines in *Nhe3*$^{-/-}$ mice, tg*Nhe3*$^{-/-}$, failed to rescue the structural and absorptive defects of global NHE3 deletion, with basal blood pressure being similar to those of *Nhe3*$^{-/-}$ mice [195, 196]. These abnormal phenotypes have been confirmed by us recently [21–23].

However, these studies using either *Nhe3*$^{-/-}$ or tg*Nhe3*$^{-/-}$ mice are unable to determine the roles of NHE3 in the proximal tubules of the kidney, since NHE3 is abundantly expressed not only in the proximal tubules of the kidney, but also in small intestines of the gut. To overcome this limitation, we have generated mutant mice with deletion of NHE3 selectively in the proximal tubules of the kidney, PT-*Nhe3*$^{-/-}$, using the state of the art Sglt2-Cre/LoxP approach [23]. We directly tested the hypothesis that deletion of NHE3 selectively in the proximal tubules of the kidney would lower basal blood pressure by inhibiting proximal tubule Na^+ reabsorption and increasing the pressure natriuresis response in mice [23]. We demonstrated that under basal conditions, PT-*Nhe3*$^{-/-}$ mice had significantly lower systolic, diastolic, and mean arterial blood pressure than WT mice, accompanied by significantly greater diuretic and natriuretic responses than WT mice, without altering 24 h fecal Na^+ excretion, plasma pH, Na^+, and bicarbonate levels. Furthermore, we demonstrated that the pressure-natriuresis response, as well natriuretic

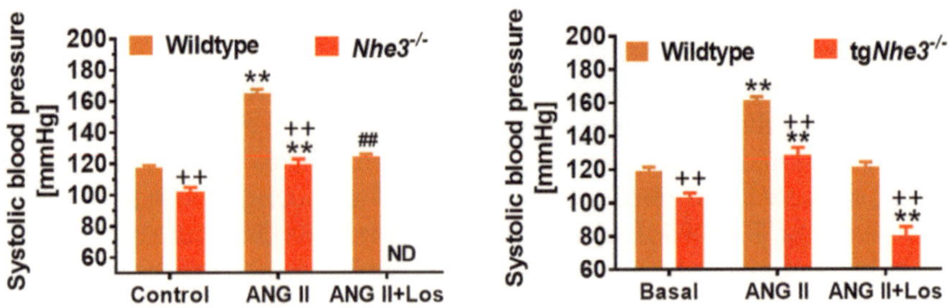

Figure 5.
Global (Nhe3$^{-/-}$) or "kidney-selective" deletion of the Na^+/H^+ exchanger 3 (NHE3) (tgNhe3$^{-/-}$) in mice significantly attenuates systolic blood pressure response to angiotensin II infusion for 2 weeks (ANG II), 1.5 mg/kg/day, i.p. $^{**}p < 0.01$ versus their control or basal; $^{++}p < 0.01$ versus wildtype; $^{\#\#}p < 0.01$ versus ANG II.

responses to acute volume expansion and a high salt diet, were significantly augmented in PT-*Nhe3*$^{-/-}$ mice [23]. Thus, our data support the scientific premise and physiological relevance that NHE3 in the proximal tubules plays an important role in maintaining basal blood pressure homeostasis, and genetic deletion of NHE3 selectively in the proximal tubules of the kidney lowers blood pressure by increasing the pressure-natriuretic response.

Recently, we further investigated whether NHE3 in small intestines and proximal tubules of the kidney plays a key role in ANG II-induced hypertension using *Nhe3*$^{-/-}$, tg*Nhe3*$^{-/-}$, and PT-*Nhe3*$^{-/-}$ mice [21, 22]. As expected, infusion of a pressor dose of ANG II, 1.5 mg/kg/day, i.p., via an osmotic minipump for 2 weeks markedly increased blood pressure and caused hypertension in C57BL/6J mice (**Figure 5**) [21, 22]. These hypertensive responses were significantly attenuated in conscious and anesthetized *Nhe3*$^{-/-}$, tg*Nhe3*$^{-/-}$, and PT-*Nhe3*$^{-/-}$ mice [21, 22, 197]. These results strongly support an important role of NHE3 not only in small intestines, but also in the proximal tubules of the kidney in maintaining basal blood pressure homeostasis and in the development of ANG II-induced hypertension.

9. Future perspectives and conclusions

Taken together, there is accumulating evidence to support the existence of the circulating (endocrine), local intratubular (paracrine), and intracellular RAS system in the kidney, especially in the proximal tubules. All major components of the RAS, including the substrate angiotensinogen, renin, ACE, ANG II, AT_1 and AT_2 receptors, have been localized in the circulation, the kidney, and in the proximal tubule. The roles of the circulating and intratubular RAS in the cardiovascular and kidney, and blood pressure regulation have been extensively studied using molecular, cellular, genetic and pharmacological approaches. It is now well-understood that AGT, prorenin, renin, ACE, ANG II and AT_1 and AT_2 receptors are not only expressed and localized in the proximal tubules under physiological conditions, but the levels of intratubular angiotensinogen, renin, ACE, and ANG II proteins are also significantly increased in the kidney in response to ANG II infusion in spite of suppression of the circulating RAS. Furthermore, there is also increasing evidence supporting the genomic roles of intracellular and nuclear ANG II in the regulation of proximal tubule reabsorption, blood pressure and the development of hypertension. Future studies should focus more on the long-term genomic and hypertensive roles of intracellular, mitochondrial and nuclear ANG II and the underlying signaling mechanisms in ANG II-dependent hypertension and target organ injury.

Acknowledgements

This work was supported in part by NIH grants, 2R01DK102429-03A1, 2R01DK067299-10A1, and 1R56HL130988-01 to Dr. Zhuo. Ana Paula de Oliveira Leite was supported by scholarships from the Ministry of Education, Brazilian Federal Agency for Support and Evaluation of Graduate Education—CAPES, and Hospital do Rim, Sao Paulo, Brazil, respectively. Drs. Chunling Zhao, Xiaowen Zheng, Jianfeng Zhang were visiting scholars from the Department of Emergency Medicine, The 2nd Affiliated Hospital, Guangxi Medical University, Nanning, Guangxi, China.

Conflict of interest

The authors declare no conflict of interest.

Author details

Xiao C. Li†, Ana Paula de Oliveira Leite†, Xu Chen, Chunling Zhao, Xiaowen Zheng, Jianfeng Zhang and Jia L. Zhuo*
Laboratory of Receptor and Signal Transduction, Department of Pharmacology and Toxicology, Division of Nephrology, Internal Medicine, University of Mississippi Medical Center, Jackson, Mississippi, USA

*Address all correspondence to: jzhuo@umc.edu

† These authors contributed equally.

References

[1] Whelton PK, Carey RM, Aronow WS, Casey DE Jr, Collins KJ, Dennison HC, et al. 2017ACC/AHA/AAPA/ABC/ACPM/AGS/APhA/ASH/ASPC/NMA/PCNA guideline for the prevention, detection, evaluation, and management of high blood pressure in adults: Executive summary: A report of the American College Of Cardiology/American Heart Association task force on clinical practice guidelines. Circulation. 2018;**138**(17):e426-e483

[2] Carey RM, Whelton PK. Prevention, detection, evaluation, and management of high blood pressure in adults: Synopsis of the 2017 American College of Cardiology/American Heart Association Hypertension Guideline. Annals of Internal Medicine. 2018;**168**(5):351-358

[3] Chobanian AV, Bakris GL, Black HR, Cushman WC, Green LA, Izzo JL Jr, et al. The seventh report of the Joint National Committee on Prevention, Detection, Evaluation, and Treatment of High Blood Pressure. Hypertension. 2003;**42**(6):1206-1252

[4] Lloyd-Jones D, Adams RJ, Brown TM, Carnethon M, Dai S, De SG, et al. Heart disease and stroke statistics—2010 update: A report from the American Heart Association. Circulation. 2010;**121**(7):e46-e215

[5] Calhoun DA, Jones D, Textor S, Goff DC, Murphy TP, Toto RD, et al. Resistant hypertension: Diagnosis, evaluation, and treatment. A scientific statement from the American Heart Association Professional Education Committee of the Council for High Blood Pressure Research. Hypertension. 2008;**51**(6):1403-1419

[6] Sacks FM, Svetkey LP, Vollmer WM, Appel LJ, Bray GA, Harsha D, et al. Effects on blood pressure of reduced dietary sodium and the Dietary Approaches to Stop Hypertension (DASH) diet. DASH-Sodium Collaborative Research Group. The New England Journal of Medicine. 2001;**344**(1):3-10

[7] Bomback AS, Toto R. Dual blockade of the renin-angiotensin-aldosterone system: Beyond the ACE inhibitor and angiotensin-II receptor blocker combination. American Journal of Hypertension. 2009;**22**(10):1032-1040

[8] Jorde UP, Ennezat PV, Lisker J, Suryadevara V, Infeld J, Cukon S, et al. Maximally recommended doses of angiotensin-converting enzyme (ACE) inhibitors do not completely prevent ACE-mediated formation of angiotensin II in chronic heart failure. Circulation. 2000;**101**(8):844-846

[9] Cook JL, Zhang Z, Re RN. In vitro evidence for an intracellular site of angiotensin action. Circulation Research. 2001;**89**(12):1138-1146

[10] Li XC, Zhu D, Zheng X, Zhang J, Zhuo JL. Intratubular and intracellular renin-angiotensin system in the kidney: A unifying perspective in blood pressure control. Clinical Science (London, England). 2018;**132**(13):1383-1401

[11] Kumar R, Singh VP, Baker KM. The intracellular renin-angiotensin system: Implications in cardiovascular remodeling. Current Opinion in Nephrology and Hypertension. 2008; **17**(2):168-173

[12] Murphy JE, Padilla BE, Hasdemir B, Cottrell GS, Bunnett NW. Endosomes: A legitimate platform for the signaling train. Proceedings of the National Academy of Sciences of the United States of America. 2009;**0906541106**:1-8

[13] Cottrell GS, Padilla BE, Amadesi S, Poole DP, Murphy JE, Hardt M, et al. Endosomal endothelin-converting enzyme-1: A regulator of beta-arrestin-dependent ERK signaling. The Journal

of Biological Chemistry. 2009;**284**(33): 22411-22425

[14] Li XC, Carretero OA, Zhuo JL. Angiotensin II AT_{1a} receptor siRNA inhibits receptor-mediated angiotensin II endocytosis and NHE3 expression in proximal tubule cells. Journal of the American Society of Nephrology. 2005; **16**:573A

[15] Zhuo JL, Li XC, Garvin JL, Navar LG, Carretero OA. Intracellular angiotensin II induces cytosolic Ca^{2+} mobilization by stimulating intracellular AT_1 receptors in proximal tubule cells. American Journal of Physiology. Renal Physiology. 2006; **290**:F1382-F1390

[16] Li XC, Hopfer U, Zhuo JL. Novel signaling mechanisms of intracellular angiotensin II-induced NHE3 expression and activation in mouse proximal tubule cells. American Journal of Physiology. Renal Physiology. 2012; **303**(12):F1617-F1628

[17] Li XC, Cook JL, Rubera I, Tauc M, Zhang F, Zhuo JL. Intrarenal transfer of an intracellular cyan fluorescent fusion of angiotensin II selectively in proximal tubules increases blood pressure in rats and mice. American Journal of Physiology. Renal Physiology. 2011;**300**: F1076-F1088

[18] Li XC, Zhuo JL. Proximal tubule-dominant transfer of AT_{1a} receptors induces blood pressure responses to intracellular angiotensin II in AT_{1a} receptor-deficient mice. American Journal of Physiology. Regulatory, Integrative and Comparative Physiology. 2013;**304**:R588-R598

[19] Li XC, Navar LG, Shao Y, Zhuo JL. Genetic deletion of AT_{1a} receptors attenuates intracellular accumulation of angiotensin II in the kidney of AT_{1a} receptor-deficient mice. American Journal of Physiology. Renal Physiology. 2007;**293**:F586-F593

[20] Li XC, Zhuo JL. In vivo regulation of AT_{1a} receptor-mediated intracellular uptake of [^{125}I]-Val^5-angiotensin II in the kidneys and adrenal glands of AT_{1a} receptor-deficient mice. American Journal of Physiology. Renal Physiology. 2008;**294**:F293-F302

[21] Li XC, Shull GE, Miguel-Qin E, Zhuo JL. Role of the Na^+/H^+ exchanger 3 in angiotensin II-induced hypertension. Physiological Genomics. 2015;**47**(10): 479-487

[22] Li XC, Shull GE, Miguel-Qin E, Chen F, Zhuo JL. Role of the Na^+/H^+ exchanger 3 in angiotensin II-induced hypertension in NHE3-deficient mice with transgenic rescue of NHE3 in small intestines. Physiological Reports. 2015; **3**(11):e12605

[23] Li XC, Soleimani M, Zhu D, Rubera I, Tauc M, Zheng X, et al. Proximal tubule-specific deletion of the NHE3 (Na^+/H^+ exchanger 3) promotes the pressure-natriuresis response and lowers blood pressure in mice. Hypertension. 2018;**72**(6):1328-1336

[24] Tewksbury DA, Frome WL, Dumas ML. Characterization of human angiotensinogen. The Journal of Biological Chemistry. 1978;**253**(11): 3817-3820

[25] Tewksbury DA, Dart RA, Travis J. The amino terminal amino acid sequence of human angiotensinogen. Biochemical and Biophysical Research Communications. 1981;**99**(4): 1311-1315

[26] Bouhnik J, Clauser E, Strosberg D, Frenoy JP, Menard J, Corvol P. Rat angiotensinogen and des(angiotensin I) angiotensinogen: Purification, characterization, and partial sequencing. Biochemistry. 1981;**20**(24): 7010-7015

[27] Clouston WM, Evans BA, Haralambidis J, Richards RI. Molecular

cloning of the mouse angiotensinogen gene. Genomics. 1988;**2**(3):240-248

[28] Kobori H, Nangaku M, Navar LG, Nishiyama A. The intrarenal renin-angiotensin system: From physiology to the pathobiology of hypertension and kidney disease. Pharmacological Reviews. 2007;**59**(3): 251-287

[29] Taugner R, Hackenthal E, Rix E, Nobiling R, Poulsen K. Immunocytochemistry of the renin-angiotensin system: Renin, angiotensinogen, angiotensin I, angiotensin II, and converting enzyme in the kidneys of mice, rats, and tree shrews. Kidney International Supplements. 1982;**12**:S33-S43

[30] Darby IA, Sernia C. In situ hybridization and immunohistochemistry of renal angiotensinogen in neonatal and adult rat kidneys. Cell and Tissue Research. 1995;**281**(2):197-206

[31] Kobori H, Harrison-Bernard LM, Navar LG. Expression of angiotensinogen mRNA and protein in angiotensin II-dependent hypertension. Journal of the American Society of Nephrology. 2001;**12**(3):431-439

[32] Kobori H, Harrison-Bernard LM, Navar LG. Urinary excretion of angiotensinogen reflects intrarenal angiotensinogen production. Kidney International. 2002;**61**(2): 579-585

[33] Kobori H, Harrison-Bernard LM, Navar LG. Enhancement of angiotensinogen expression in angiotensin II-dependent hypertension. Hypertension. 2001;**37**(5):1329-1335

[34] Matsusaka T, Niimura F, Shimizu A, Pastan I, Saito A, Kobori H, et al. Liver angiotensinogen is the primary source of renal angiotensin II. Journal of the American Society of Nephrology. 2012; **23**(7):1181-1189

[35] Tigerstedt R, Bergman PG. Niere und Kreislauf. Skandinavisches Archiv Für Physiologie. 1898;**8**:223-271

[36] Imai T, Miyazaki H, Hirose S, Hori H, Hayashi T, Kageyama R, et al. Cloning and sequence analysis of cDNA for human renin precursor. Proceedings of the National Academy of Sciences of the United States of America. 1983; **80**(24):7405-7409

[37] Taugner R, Hackenthal E, Nobiling R, Harlacher M, Reb G. The distribution of renin in the different segments of the renal arterial tree: Immunocytochemical investigation in the mouse kidney. Histochemistry. 1981;**73**(1):75-88

[38] Celio MR, Inagami T. Renin in the human kidney. Immunohistochemical localization. Histochemistry. 1981;**72**(1): 1-10

[39] Faraggiana T, Gresik E, Tanaka T, Inagami T, Lupo A. Immunohistochemical localization of renin in the human kidney. The Journal of Histochemistry and Cytochemistry. 1982;**30**(5):459-465

[40] Song K, Zhuo JL, Chai SY, Mendelsohn FA. A new method to localize active renin in tissues by autoradiography: Application to dog kidney. Kidney International. 1992; **42**(3):639-646

[41] Zhuo JL, Anderson WP, Song K, Mendelsohn FA. Autoradiographic localization of active renin in the juxtaglomerular apparatus of the dog kidney: Effects of sodium intake. Clinical and Experimental Pharmacology & Physiology. 1996; **23**(4):291-298

[42] Zhuo JL, Song K, Chai SY, Mendelsohn FA. Anatomical localization of components of the renin-angiotensin system in different organs and tissues. In: MacGregor GA, Sever PS, editors. Inhibition of the Renin-Angiotensin System: Recent Advances. Hong Kong:

Gardiner-Calwell Communications (Pacific) Ltd.; 1993

[43] Moe OW, Ujiie K, Star RA, Miller RT, Widell J, Alpern RJ, et al. Renin expression in renal proximal tubule. The Journal of Clinical Investigation. 1993; **91**(3):774-779

[44] Chen M, Harris MP, Rose D, Smart A, He XR, Kretzler M, et al. Renin and renin mRNA in proximal tubules of the rat kidney. Journal of Clinical Investigation. 1994;**94**(1):237-243

[45] Tang SS, Jung F, Diamant D, Brown D, Bachinsky D, Hellman P, et al. Temperature-sensitive SV40 immortalized rat proximal tubule cell line has functional renin-angiotensin system. The American Journal of Physiology. 1995;**268**(3 Pt 2):F435-F446

[46] Taugner R, Hackenthal E, Inagami T, Nobiling R, Poulsen K. Vascular and tubular renin in the kidneys of mice. Histochemistry. 1982; **75**(4):473-484

[47] Iwao H, Nakamura N, Ikemoto F, Yamamoto K. Subcellular localization of exogenously administered renin in mouse kidney. Japanese Circulation Journal. 1983;**47**(10):1198-1202

[48] Soubrier F, Alhenc-Gelas F, Hubert C, Allegrini J, John M, Tregear G, et al. Two putative active centers in human angiotensin I-converting enzyme revealed by molecular cloning. Proceedings of the National Academy of Sciences of the United States of America. 1988;**85**(24):9386-9390

[49] Bernstein KE, Martin BM, Bernstein EA, Linton J, Striker L, Striker G. The isolation of angiotensin-converting enzyme cDNA. The Journal of Biological Chemistry. 1988;**263**(23):11021-11024

[50] Esther CR, Marino EM, Howard TE, Machaud A, Corvol P, Capecchi MR, et al. The critical role of tissue angiotensin-converting enzyme as revealed by gene targeting in mice. The Journal of Clinical Investigation. 1997; **99**:2375-2385

[51] Bernstein KE. Views of the renin-angiotensin system: Brilling, mimsy, and slithy tove. Hypertension. 2006; **47**(3):509-514

[52] Bernstein KE, Ong FS, Blackwell WL, Shah KH, Giani JF, Gonzalez-Villalobos RA, et al. A modern understanding of the traditional and nontraditional biological functions of angiotensin-converting enzyme. Pharmacological Reviews. 2012;**65**(1): 1-46

[53] Chai SY, Allen AM, Adam WR, Mendelsohn FA. Local actions of angiotensin II: Quantitative in vitro autoradiographic localization of angiotensin II receptor binding and angiotensin converting enzyme in target tissues. Journal of Cardiovascular Pharmacology. 1986;**8**(Suppl 10): S35-S39

[54] Harrison-Bernard LM, Zhuo JL, Kobori H, Ohishi M, Navar LG. Intrarenal AT_1 receptor and ACE binding in ANG II-induced hypertensive rats. American Journal of Physiology. Renal Physiology. 2002; **282**(1):F19-F25

[55] Bruneval P, Hinglais N, Alhenc-Gelas F, Tricottet V, Corvol P, Menard J, et al. Angiotensin I converting enzyme in human intestine and kidney. Ultrastructural immunohistochemical localization. Histochemistry. 1986;**85**(1): 73-80

[56] Danilov SM, Faerman AI, Printseva OY, Martynov AV, Sakharov IY, Trakht IN. Immunohistochemical study of angiotensin-converting enzyme in human tissues using monoclonal antibodies. Histochemistry. 1987;**87**(5): 487-490

[57] Schulz WW, Hagler HK, Buja LM, Erdos EG. Ultrastructural localization of angiotensin I-converting enzyme (EC 3.4.15.1) and neutral metalloendopeptidase (EC 3.4.24.11) in the proximal tubule of the human kidney. Laboratory Investigation. 1988; **59**(6):789-797

[58] Navar LG, Kobori H, Prieto-Carrasquero M. Intrarenal angiotensin II and hypertension. Current Hypertension Reports. 2003;**5**(2): 135-143

[59] Navar LG, Carmines PK, Huang WC, Mitchell KD. The tubular effects of angiotensin II. Kidney International. Supplement. 1987;**20**:S81-S88

[60] Harris PJ, Navar LG. Tubular transport responses to angiotensin II. American Journal of Physiology. Renal Physiology. 1985;**248**:F621-F630

[61] Crowley SD, Gurley SB, Herrera MJ, Ruiz P, Griffiths R, Kumar AP, et al. Angiotensin II causes hypertension and cardiac hypertrophy through its receptors in the kidney. Proceedings of the National Academy of Sciences of the United States of America. 2006;**103**(47): 17985-17990

[62] Crowley SD, Gurley SB, Oliverio MI, Pazmino AK, Griffiths R, Flannery PJ, et al. Distinct roles for the kidney and systemic tissues in blood pressure regulation by the renin-angiotensin system. The Journal of Clinical Investigation. 2005;**115**(4):1092-1099

[63] Carey RM, Siragy HM. Newly recognized components of the renin-angiotensin system: Potential roles in cardiovascular and renal regulation. Endocrine Reviews. 2003;**24**(3):261-271

[64] Navar LG, Kobori H, Prieto MC, Gonzalez-Villalobos RA. Intratubular renin-angiotensin system in hypertension. Hypertension. 2011;**57**(3): 355-362

[65] Nishiyama A, Seth DM, Navar LG. Renal interstitial fluid concentrations of angiotensins I and II in anesthetized rats. Hypertension. 2002;**39**(1):129-134

[66] Siragy HM, Howell NL, Ragsdale NV, Carey RM. Renal interstitial fluid angiotensin. Modulation by anesthesia, epinephrine, sodium depletion, and renin inhibition. Hypertension. 1995; **25**(5):1021-1024

[67] Zhuo JL, Imig JD, Hammond TG, Orengo S, Benes E, Navar LG. Ang II accumulation in rat renal endosomes during Ang II-induced hypertension: Role of AT_1 receptor. Hypertension. 2002;**39**(1):116-121

[68] Shaltout HA, Westwood BM, Averill DB, Ferrario CM, Figueroa JP, Diz DI, et al. Angiotensin metabolism in renal proximal tubules, urine, and serum of sheep: Evidence for ACE2-dependent processing of angiotensin II. American Journal of Physiology. Renal Physiology. 2007;**292**(1):F82-F91

[69] Prieto-Carrasquero MC, Kobori H, Ozawa Y, Gutierrez A, Seth D, Navar LG. AT1 receptor-mediated enhancement of collecting duct renin in angiotensin II-dependent hypertensive rats. American Journal of Physiology. Renal Physiology. 2005;**289**(3):F632-F637

[70] Zou LX, Hymel A, Imig JD, Navar LG. Renal accumulation of circulating angiotensin II in angiotensin II-infused rats. Hypertension. 1996;**27**(3 Pt 2): 658-662

[71] Chappell MC. Nonclassical renin-angiotensin system and renal function. Comprehensive Physiology. 2012;**2**(4): 2733-2752

[72] Zhuo JL, Li XC. New insights and perspectives on intrarenal renin-angiotensin system: Focus on intracrine/intracellular angiotensin II. Peptides. 2011;**32**(7):1551-1565

[73] Zou LX, Imig JD, von Thun AM, Hymel A, Ono H, Navar LG. Receptor-mediated intrarenal angiotensin II augmentation in angiotensin II-infused rats. Hypertension. 1996;**28**(4):669-677

[74] Imig JD, Navar GL, Zou LX, O'Reilly KC, Allen PL, Kaysen JH, et al. Renal endosomes contain angiotensin peptides, converting enzyme, and AT_{1a} receptors. The American Journal of Physiology. 1999;**277**(2 Pt 2):F303-F311

[75] Abadir PM, Foster DB, Crow M, Cooke CA, Rucker JJ, Jain A, et al. Identification and characterization of a functional mitochondrial angiotensin system. Proceedings of the National Academy of Sciences of the United States of America. 2011;**108**(36): 14849-14854

[76] Timmermans PB, Wong PC, Chiu AT, Herblin WF, Benfield P, Carini DJ, et al. Angiotensin II receptors and angiotensin II receptor antagonists. Pharmacological Reviews. 1993;**45**(2): 205-251

[77] de Gasparo M, Catt KJ, Inagami T, Wright JW, Unger T. International union of pharmacology. XXIII. The angiotensin II receptors. Pharmacological Reviews. 2000;**52**(3): 415-472

[78] Karnik SS, Unal H, Kemp JR, Tirupula KC, Eguchi S, Vanderheyden PM, et al. International union of basic and clinical pharmacology. XCIX. Angiotensin receptors: Interpreters of pathophysiological angiotensinergic stimuli. Pharmacological Reviews. 2015; **67**(4):754-819

[79] Murphy TJ, Alexander RW, Griendling KK, Runge MS, Bernstein KE. Isolation of a cDNA encoding the vascular type-1 angiotensin II receptor. Nature. 1991;**16**(351):233-236

[80] Sasaki K, Yamano Y, Bardhan S, Iwai N, Murray JJ, Hasegawa M, et al. Cloning and expression of a complementary DNA encoding a bovine adrenal angiotensin II type-1 receptor. Nature. 1991;**351**(6323):230-233

[81] Zhuo JL, Ferrao FM, Zheng Y, Li XC. New frontiers in the intrarenal renin-angiotensin system: A critical review of classical and new paradigms. Frontiers in Endocrinology (Lausanne). 2013;**4**:166

[82] Mukoyama M, Nakajima M, Horiuchi M, Sasamura H, Pratt RE, Dzau VJ. Expression cloning of type 2 angiotensin II receptor reveals a unique class of seven-transmembrane receptors. The Journal of Biological Chemistry. 1993;**268**(33):24539-24542

[83] Nakajima M, Mukoyama M, Pratt RE, Horiuchi M, Dzau VJ. Cloning of cDNA and analysis of the gene for mouse angiotensin II type 2 receptor. Biochemical and Biophysical Research Communications. 1993;**197**(2):393-399

[84] Kambayashi Y, Bardhan S, Takahashi K, Tsuzuki S, Inui H, Hamakubo T, et al. Molecular cloning of a novel angiotensin II receptor isoform involved in phosphotyrosine phosphatase inhibition. The Journal of Biological Chemistry. 1993;**268**(33): 24543-24546

[85] Zhuo JL, Alcorn D, Harris PJ, Mendelsohn FA. Localization and properties of angiotensin II receptors in rat kidney. Kidney International. Supplement. 1993;**42**:S40-S46

[86] Zhuo JL, Song K, Harris PJ, Mendelsohn FA. In vitro autoradiography reveals predominantly AT1 angiotensin II receptors in rat kidney. Renal Physiology and Biochemistry. 1992;**15**(5):231-239

[87] Zhuo JL, Allen AM, Alcorn D, MacGregor D, Aldred GP, Mendelsohn FA. The distribution of angiotensin II receptors. In: Laragh JH, Brenner BM,

editors. Hypertension: Pathology, Diagnosis & Management. 2nd ed. New York: Raven Press; 1995. pp. 1739-1762

[88] Gwathmey T, Shaltout HA, Pendergrass KD, Pirro NT, Figueroa JP, Rose JC, et al. Nuclear angiotensin II—type 2 (AT_2) receptors are functionally linked to nitric oxide production. American Journal of Physiology-Renal Physiology. 2009;**296**:F1484-F1493

[89] Gwathmey TM, Westwood BM, Pirro NT, Tang L, Rose JC, Diz DI, et al. Nuclear angiotensin-(1-7) receptor is functionally coupled to the formation of nitric oxide. American Journal of Physiology. Renal Physiology. 2010; **299**(5):F983-F990

[90] Wilson BA, Nautiyal M, Gwathmey TM, Rose JC, Chappell MC. Evidence for a mitochondrial angiotensin-(1-7) system in the kidney. American Journal of Physiology. Renal Physiology. 2016; **310**(7):F637-F645

[91] Zhuo JL, MacGregor D, Mendelsohn FA. Comparative distribution of angiotensin II receptor subtypes in mammalian adrenal glands. In: Vinson GP, Anderson DC, editors. Vascular, Adrenal and Hypertension. London: Journal of Endocrinology Pty Ltd.; 1995. pp. 53-68

[92] Ozono R, Wang ZQ, Moore AF, Inagami T, Siragy HM, Carey RM. Expression of the subtype 2 angiotensin (AT_2) receptor protein in rat kidney. Hypertension. 1997;**30**(5):1238-1246

[93] Kemp BA, Howell NL, Gildea JJ, Keller SR, Padia SH, Carey RM. AT(2) receptor activation induces natriuresis and lowers blood pressure. Circulation Research. 2014;**115**(3):388-399

[94] Touyz RM, Schiffrin EL. Signal transduction mechanisms mediating the physiological and pathophysiological actions of angiotensin II in vascular smooth muscle cells. Pharmacological Reviews. 2000;**52**:639-672

[95] Mehta PK, Griendling KK. Angiotensin II cell signaling: Physiological and pathological effects in the cardiovascular system. American Journal of Physiology. Cell Physiology. 2007;**292**(1):C82-C97

[96] Baker KM, Booz GW, Dostal DE. Cardiac actions of angiotensin II: Role of an intracardiac renin-angiotensin system. Annual Review of Physiology. 1992;**54**:227-241

[97] Re RN. On the biological actions of intracellular angiotensin. Hypertension. 2000;**35**(6):1189-1190

[98] Re R. Intracellular renin-angiotensin system: The tip of the intracrine physiology iceberg. American Journal of Physiology-Heart and Circulatory Physiology. 2007;**293**(2):H905-H906

[99] Kumar R, Singh VP, Baker KM. The intracellular renin-angiotensin system: A new paradigm. Trends in Endocrinology and Metabolism. 2007; **18**(5):208-214

[100] Zhuo JL, Li XC. Novel roles of intracrine angiotensin II and signalling mechanisms in kidney cells. Journal of the Renin-Angiotensin-Aldosterone System. 2007;**8**(1):23-33

[101] De Mello WC, Danser AH. Angiotensin II and the heart: On the intracrine renin-angiotensin system. Hypertension. 2000;**35**(6):1183-1188

[102] Li XC, Zhuo JL. Intracellular ANG II directly induces in vitro transcription of TGF-β1, MCP-1, and NHE-3 mRNAs in isolated rat renal cortical nuclei via activation of nuclear AT_{1a} receptors. American Journal of Physiology. Cell Physiology. 2008;**294**(4):C1034-C1045

[103] Ferrao FM, Cardoso LHD, Drummond HA, Li XC, Zhuo JL,

Gomes DS, et al. Luminal ANG II is internalized as a complex with AT_1R/AT_2R heterodimers to target endoplasmic reticulum in LLC-PK1 cells. American Journal of Physiology. Renal Physiology. 2017;**313**(2): F440-F449

[104] Brasier AR, Jamaluddin M, Han Y, Patterson C, Runge MS. Angiotensin II induces gene transcription through cell-type-dependent effects on the nuclear factor-kappaB (NF-kappaB) transcription factor. Molecular and Cellular Biochemistry. 2000;**212**(1-2): 155-169

[105] Ruiz-Ortega M, Lorenzo O, Ruperez M, Konig S, Wittig B, Egido J. Angiotensin II activates nuclear transcription factor κB through AT_1 and AT_2 in vascular smooth muscle cells: Molecular mechanisms. Circulation Research. 2000;**86**(12):1266-1272

[106] Li XC, Zhuo JL. Nuclear factor-κB as a hormonal intracellular signaling molecule: Focus on angiotensin II-induced cardiovascular and renal injury. Current Opinion in Nephrology and Hypertension. 2008;**17**(1):37-43

[107] Takahashi M, Suzuki E, Takeda R, Oba S, Nishimatsu H, Kimura K, et al. Angiotensin II and tumor necrosis factor-alpha synergistically promote monocyte chemoattractant protein-1 expression: Roles of NF-κB, p38, and reactive oxygen species. American Journal of Physiology. Heart and Circulatory Physiology. 2008;**294**(6): H2879-H2888

[108] Zhuo JL. Monocyte chemoattractant protein-1: A key mediator of angiotensin II-induced target organ damage in hypertensive heart disease? Journal of Hypertension. 2004;**22**(3):451-454

[109] Kagami S, Border WA, Miller DE, Noble NA. Angiotensin II stimulates extracellular matrix protein synthesis through induction of transforming growth factor-beta expression in rat glomerular mesangial cells. The Journal of Clinical Investigation. 1994;**93**:2431-2437

[110] Wolf G, Ziyadeh FN, Stahl RA. Angiotensin II stimulates expression of transforming growth factor beta receptor type II in cultured mouse proximal tubular cells. Journal of Molecular Medicine. 1999;**77**(7):556-564

[111] Kurtz TW, Gardner DG. Transcription-modulating drugs: A new frontier in the treatment of essential hypertension. Hypertension. 1998;**32**: 380-386

[112] Kurtz TW, Pravenec M. Molecule-specific effects of angiotensin II-receptor blockers independent of the renin-angiotensin system. American Journal of Hypertension. 2008;**21**(8): 852-859

[113] Kurtz TW. Beyond the classic angiotensin-receptor-blocker profile. Nature Clinical Practice. Cardiovascular Medicine. 2008;**5**(Suppl 1):S19-S26

[114] Conchon S, Monnot C, Teutsch B, Corvol P, Clauser E. Internalization of the rat AT_{1a} and AT_{1b} receptors: Pharmacological and functional requirements. FEBS Letters. 1994; **349**(3):365-370

[115] Schupp M, Janke J, Clasen R, Unger T, Kintscher U. Angiotensin type 1 receptor blockers induce peroxisome proliferator-activated receptor-gamma activity. Circulation. 2004;**109**(17): 2054-2057

[116] Kurtz TW, Klein U. Next generation multifunctional angiotensin receptor blockers. Hypertension Research. 2009;**32**(10):826-834

[117] Mauer M, Zinman B, Gardiner R, Suissa S, Sinaiko A, Strand T, et al. Renal and retinal effects of enalapril and losartan in type 1 diabetes. The New

England Journal of Medicine. 2009; **361**(1):40-51

[118] van Kats JP, Schalekamp MA, Verdouw PD, Duncker DJ, Danser AH. Intrarenal angiotensin II: Interstitial and cellular levels and site of production. Kidney International. 2001;**60**(6): 2311-2317

[119] von Thun AM, Vari RC, El Dahr SS, Navar LG. Augmentation of intrarenal angiotensin II levels by chronic angiotensin II infusion. The American Journal of Physiology. 1994;**266**(1 Pt 2): F120-F128

[120] Zou LX, Imig JD, Hymel A, Navar LG. Renal uptake of circulating angiotensin II in Val^5-angiotensin II infused rats is mediated by AT^1 receptor. American Journal of Hypertension. 1998;**11**(5):570-578

[121] Shao W, Seth DM, Navar LG. Augmentation of endogenous intrarenal angiotensin II levels in Val^5-Ang II infused rats. American Journal of Physiology. Renal Physiology. 2009; **296**(5):F1067-F1071

[122] Li XC, Zhuo JL. Mechanisms of AT_{1a} receptor-mediated uptake of angiotensin II by proximal tubule cells: A novel role of the multiligand endocytic receptor megalin. American Journal of Physiology. Renal Physiology. 2014;**307**(2):F222-F233

[123] Li XC, Gu V, Miguel-Qin E, Zhuo JL. Role of caveolin 1 in AT_{1a} receptor-mediated uptake of angiotensin II in the proximal tubule of the kidney. American Journal of Physiology. Renal Physiology. 2014;**307**(8):F949-F961

[124] Li XC, Carretero OA, Navar LG, Zhuo JL. AT_1 receptor-mediated accumulation of extracellular angiotensin II in proximal tubule cells: Role of cytoskeleton microtubules and tyrosine phosphatases. American Journal of Physiology. Renal Physiology. 2006;**291**:F375-F383

[125] Li XC, Zhuo JL. Selective knockdown of AT_1 receptors by RNA interference inhibits Val^5-Ang II endocytosis and NHE-3 expression in immortalized rabbit proximal tubule cells. American Journal of Physiology. Cell Physiology. 2007;**293**:C367-C378

[126] Zhuo JL, Carretero OA, Li XC. Effects of AT_1 receptor-mediated endocytosis of extracellular Ang II on activation of nuclear factor-κB in proximal tubule cells. Annals of the New York Academy of Sciences. 2006;**1091**: 336-345

[127] Brown GP, Douglas JG. Angiotensin II binding sites on isolated rat renal brush border membranes. Endocrinology. 1982;**111**(6):1830-1836

[128] Douglas JG. Angiotensin receptor subtypes of the kidney cortex. The American Journal of Physiology. 1987; **253**(1 Pt 2):F1-F7

[129] Dulin NO, Ernsberger P, Suciu DJ, Douglas JG. Rabbit renal epithelial angiotensin II receptors. The American Journal of Physiology. 1994;**267**(5 Pt 2): F776-F782

[130] Becker BN, Cheng HF, Burns KD, Harris RC. Polarized rabbit type 1 angiotensin II receptors manifest differential rates of endocytosis and recycling. The American Journal of Physiology. 1995;**269**(4 Pt 1):C1048-C1056

[131] Thekkumkara TJ, Cookson R, Linas SL. Angiotensin (AT_{1a}) receptor-mediated increases in transcellular sodium transport in proximal tubule cells. The American Journal of Physiology. 1998;**274**:F897-F905

[132] Christensen EI, Birn H. Megalin and cubilin: Synergistic endocytic receptors in renal proximal tubule.

American Journal of Physiology. Renal Physiology. 2001;**280**(4):F562-F573

[133] Zhai XY, Nielsen R, Birn H, Drumm K, Mildenberger S, Freudinger R, et al. Cubilin- and megalin-mediated uptake of albumin in cultured proximal tubule cells of opossum kidney. Kidney International. 2000;**58**(4):1523-1533

[134] Kozyraki R, Fyfe J, Verroust PJ, Jacobsen C, Dautry-Varsat A, Gburek J, et al. Megalin-dependent cubilin-mediated endocytosis is a major pathway for the apical uptake of transferrin in polarized epithelia. Proceedings of the National Academy of Sciences of the United States of America. 2001;**98**(22):12491-12496

[135] Leheste JR, Rolinski B, Vorum H, Hilpert J, Nykjaer A, Jacobsen C, et al. Megalin knockout mice as an animal model of low molecular weight proteinuria. The American Journal of Pathology. 1999;**155**(4):1361-1370

[136] Gonzalez-Villalobos R, Klassen RB, Allen PL, Navar LG, Hammond TG. Megalin binds and internalizes angiotensin II. American Journal of Physiology. Renal Physiology. 2005;**288**: F420-F427

[137] Anborgh PH, Seachrist JL, Dale LB, Ferguson SS. Receptor/beta-arrestin complex formation and the differential trafficking and resensitization of β2-adrenergic and angiotensin II type 1A receptors. Molecular Endocrinology. 2000;**14**(12):2040-2053

[138] Ferguson SS. Evolving concepts in G protein-coupled receptor endocytosis: The role in receptor desensitization and signaling. Pharmacological Reviews. 2001;**53**(1):1-24

[139] Zhang J, Barak LS, Anborgh PH, Laporte SA, Caron MG, Ferguson SS. Cellular trafficking of G protein-coupled receptor/beta-arrestin endocytic complexes. The Journal of Biological Chemistry. 1999;**274**(16): 10999-11006

[140] Thomas WG, Thekkumkara TJ, Baker KM. Molecular mechanisms of angiotensin II (AT1A) receptor endocytosis. Clinical and Experimental Pharmacology & Physiology. Supplement. 1996;**3**:S74-S80

[141] Rajagopal K, Whalen EJ, Violin JD, Stiber JA, Rosenberg PB, Premont RT, et al. Beta-arrestin2-mediated inotropic effects of the angiotensin II type 1A receptor in isolated cardiac myocytes. Proceedings of the National Academy of Sciences of the United States of America. 2006;**103**(44):16284-16289

[142] Rappoport JZ, Kemal S, Benmerah A, Simon SM. Dynamics of clathrin and adaptor proteins during endocytosis. American Journal of Physiology. Cell Physiology. 2006;**291**(5):C1072-C1081

[143] Zhang J, Ferguson SS, Barak LS, Menard L, Caron MG. Dynamin and beta-arrestin reveal distinct mechanisms for G protein-coupled receptor internalization. The Journal of Biological Chemistry. 1996;**271**(31):18302-18305

[144] Qian H, Pipolo L, Thomas WG. Association of beta-Arrestin 1 with the type 1A angiotensin II receptor involves phosphorylation of the receptor carboxyl terminus and correlates with receptor internalization. Molecular Endocrinology. 2001;**15**(10):1706-1719

[145] Seachrist JL, Laporte SA, Dale LB, Babwah AV, Caron MG, Anborgh PH, et al. Rab5 association with the angiotensin II type 1A receptor promotes Rab5 GTP binding and vesicular fusion. The Journal of Biological Chemistry. 2002;**277**(1): 679-685

[146] Dale LB, Seachrist JL, Babwah AV, Ferguson SS. Regulation of angiotensin II type 1A receptor intracellular retention, degradation, and recycling by

Rab5, Rab7, and Rab11 GTPases. The Journal of Biological Chemistry. 2004; **279**(13):13110-13118

[147] Becker BN, Cheng HF, Harris RC. Apical ANG II-stimulated PLA2 activity and Na^+ flux: A potential role for Ca^{2+}-independent PLA. The American Journal of Physiology. 1997;**273**(4 Pt 2): F554-F562

[148] Elkjaer ML, Birn H, Agre P, Christensen EI, Nielsen S. Effects of microtubule disruption on endocytosis, membrane recycling and polarized distribution of Aquaporin-1 and gp330 in proximal tubule cells. European Journal of Cell Biology. 1995;**67**(1):57-72

[149] Schelling JR, Linas SL. Angiotensin II-dependent proximal tubule sodium transport requires receptor-mediated endocytosis. The American Journal of Physiology. 1994;**266**(3 Pt 1):C669-C675

[150] Schelling JR, Hanson AS, Marzec R, Linas SL. Cytoskeleton-dependent endocytosis is required for apical type 1 angiotensin II receptor-mediated phospholipase C activation in cultured rat proximal tubule cells. The Journal of Clinical Investigation. 1992;**90**(6): 2472-2480

[151] Li XC, Hopfer U, Zhuo JL. AT_1 receptor-mediated uptake of angiotensin II and NHE-3 expression in proximal tubule cells through the microtubule-dependent endocytic pathway. American Journal of Physiology. Renal Physiology. 2009; **297**(5):F1342-F1352

[152] Robertson A, Khairallah P. Angiotensin II: Rapid localization in nuclei of smooth and cardiac muscle. Science. 1971;**172**:1138-1139

[153] Bianchi C, Gutkowska J, De Lean A, Ballak M, Anand-Srivastava MB, Genest J, et al. Fate of [125I]angiotensin II in adrenal zona glomerulosa cells. Endocrinology. 1986;**118**:2605-2607

[154] Cook JL, Mills SJ, Naquin RT, Alam J, Re RN. Cleavage of the angiotensin II type 1 receptor and nuclear accumulation of the cytoplasmic carboxy-terminal fragment. American Journal of Physiology. Cell Physiology. 2007;**292**(4):C1313-C1322

[155] Chen R, Mukhin YV, Garnovskaya MN, Thielen TE, Iijima Y, Huang C, et al. A functional angiotensin II receptor-GFP fusion protein: Evidence for agonist-dependent nuclear translocation. American Journal of Physiology. Renal Physiology. 2000; **279**(3):F440-F448

[156] Morinelli TA, Raymond JR, Baldys A, Yang Q, Lee MH, Luttrell L, et al. Identification of a putative nuclear localization sequence within the angiotensin II AT_{1a} receptor associated with nuclear activation. American Journal of Physiology. Cell Physiology. 2007;**292**(4):C1398-C1408

[157] du Cheyron D, Chalumeau C, Defontaine N, Klein C, Kellermann O, Paillard M, et al. Angiotensin II stimulates NHE3 activity by exocytic insertion of the transporter: Role of PI 3-kinase. Kidney International. 2003; **64**(3):939-949

[158] Liu FY, Cogan MG. Angiotensin II stimulation of hydrogen ion secretion in the rat early proximal tubule. Modes of action, mechanism, and kinetics. Journal of Clinical Investigation 1988;**82**(2): 601-607

[159] Bloch RD, Zikos D, Fisher KA, Schleicher L, Oyama M, Cheng JC, et al. Activation of proximal tubular Na^+-H^+ exchange by angiotensin II. The American Journal of Physiology. 1992; **263**(1 Pt 2):F135-F143

[160] Geibel J, Giebisch G, Boron WF. Angiotensin II stimulates both Na^+-H^+ exchange and Na^+/HCO_3^- cotransport in the rabbit proximal tubule. Proceedings of the National Academy of

Sciences of the United States of America. 1990;**87**(20):7917-7920

[161] Reilly AM, Harris PJ, Williams DA. Biphasic effect of angiotensin II on intracellular sodium concentration in rat proximal tubules. The American Journal of Physiology. 1995;**269**(3 Pt 2):F374-F380

[162] Jourdain M, Amiel C, Friedlander G. Modulation of Na^+-H^+ exchange activity by angiotensin II in opossum kidney cells. The American Journal of Physiology. 1992;**263**(6 Pt 1):C1141-C1146

[163] Houillier P, Chambrey R, Achard JM, Froissart M, Poggioli J, Paillard M. Signaling pathways in the biphasic effect of angiotensin II on apical Na^+-H^+ antiport activity in proximal tubule. Kidney International. 1996;**50**(5): 1496-1505

[164] Han HJ, Park SH, Koh HJ, Taub M. Mechanism of regulation of Na^+ transport by angiotensin II in primary renal cells. Kidney International. 2000; **57**(6):2457-2467

[165] Tsuganezawa H, Preisig PA, Alpern RJ. Dominant negative c-Src inhibits angiotensin II induced activation of NHE3 in OKP cells. Kidney International. 1998;**54**(2):394-398

[166] Du Z, Ferguson W, Wang T. Role of PKC and calcium in modulation of effects of angiotensin II on sodium transport in proximal tubule. American Journal of Physiology. Renal Physiology. 2003;**284**(4):F688-F692

[167] Wang T, Chan YL. The role of phosphoinositide turnover in mediating the biphasic effect of angiotensin II on renal tubular transport. The Journal of Pharmacology and Experimental Therapeutics. 1991;**256**(1):309-317

[168] Karim Z, Defontaine N, Paillard M, Poggioli J. Protein kinase C isoforms in rat kidney proximal tubule: Acute effect of angiotensin II. The American Journal of Physiology. 1995;**269**(1 Pt 1):C134-C140

[169] Schelling JR, Singh H, Marzec R, Linas SL. Angiotensin II-dependent proximal tubule sodium transport is mediated by cAMP modulation of phospholipase C. The American Journal of Physiology. 1994;**267**(5 Pt 1):C1239-C1245

[170] Lea JP, Jin SG, Roberts BR, Shuler MS, Marrero MB, Tumlin JA. Angiotensin II stimulates calcineurin activity in proximal tubule epithelia through AT-1 receptor-mediated tyrosine phosphorylation of the PLC-gamma1 isoform. Journal of the American Society of Nephrology. 2002; **13**(7):1750-1756

[171] Liu FY, Cogan MG. Angiotensin II stimulates early proximal bicarbonate absorption in the rat by decreasing cyclic adenosine monophosphate. The Journal of Clinical Investigation. 1989; **84**(1):83-91

[172] Dulin NO, Alexander LD, Harwalkar S, Falck JR, Douglas JG. Phospholipase A2-mediated activation of mitogen-activated protein kinase by angiotensin II. Proceedings of the National Academy of Sciences of the United States of America. 1998;**95**(14): 8098-8102

[173] Ruiz-Ortega M, Lorenzo O, Ruperez M, Blanco J, Egido J. Systemic infusion of angiotensin II into normal rats activates nuclear factor-kappaB and AP-1 in the kidney: Role of AT_1 and AT_2 receptors. The American Journal of Pathology. 2001;**158**(5): 1743-1756

[174] Haller H, Lindschau C, Erdmann B, Quass P, Luft FC. Effects of intracellular angiotensin II in vascular smooth muscle cells. Circulation Research. 1996; **79**(4):765-772

[175] Haller H, Lindschau C, Quass P, Luft FC. Intracellular actions of angiotensin II in vascular smooth muscle cells. Journal of the American Society of Nephrology. 1999;**10**(Suppl 11):S75-S83

[176] De Mello WC. Renin increments the inward calcium current in the failing heart. Journal of Hypertension. 2006; **24**(6):1181-1186

[177] Zhuo JL. Intracrine renin and angiotensin II: A novel role in cardiovascular and renal cellular regulation. Journal of Hypertension. 2006;**24**(6):1017-1020

[178] Baker KM, Kumar R. Intracellular angiotensin II induces cell proliferation independent of AT_1 receptor. American Journal of Physiology. Cell Physiology. 2006;**291**(5):C995-C1001

[179] Baker KM, Chernin MI, Schreiber T, Sanghi S, Haiderzaidi S, Booz GW, et al. Evidence of a novel intracrine mechanism in angiotensin II-induced cardiac hypertrophy. Regulatory Peptides. 2004;**120**(1-3):5-13

[180] van Kats JP, Methot D, Paradis P, Silversides DW, Reudelhuber TL. Use of a biological peptide pump to study chronic peptide hormone action in transgenic mice. Direct and indirect effects of angiotensin II on the heart. The Journal of Biological Chemistry. 2001;**276**(47):44012-44017

[181] Xu J, Carretero OA, Lin CX, Cavasin MA, Shesely EG, Yang JJ, et al. Role of cardiac overexpression of ANG II in the regulation of cardiac function and remodeling postmyocardial infarction. American Journal of Physiology. Heart and Circulatory Physiology. 2007;**293**(3):H1900-H1907

[182] Ding Y, Davisson RL, Hardy DO, Zhu LJ, Merrill DC, Catterall JF, et al. The kidney androgen-regulated protein promoter confers renal proximal tubule cell-specific and highly androgen-responsive expression on the human angiotensinogen gene in transgenic mice. The Journal of Biological Chemistry. 1997;**272**(44): 28142-28148

[183] Ding Y, Sigmund CD. Androgen-dependent regulation of human angiotensinogen expression in KAP-hAGT transgenic mice. American Journal of Physiology. Renal Physiology. 2001;**280**(1):F54-F60

[184] Soler M, Tornavaca O, Sole E, Menoyo A, Hardy D, Catterall JF, et al. Hormone-specific regulation of the kidney androgen-regulated gene promoter in cultured mouse renal proximal-tubule cells. The Biochemical Journal. 2002;**366**(Pt 3):757-766

[185] Meseguer A, Catterall JF. Cell-specific expression of kidney androgen-regulated protein messenger RNA is under multihormonal control. Molecular Endocrinology. 1990;**4**(8): 1240-1248

[186] Hediger MA, Rhoads DB. Molecular physiology of sodium-glucose cotransporters. Physiological Reviews. 1994;**74**(4):993-1026

[187] Weinman EJ, Shenolikar S. Regulation of the renal brush border membrane Na^+-H^+ exchanger. Annual Review of Physiology. 1993;**55**:289-304

[188] Biemesderfer D, Rutherford PA, Nagy T, Pizzonia JH, Abu-Alfa AK, Aronson PS. Monoclonal antibodies for high-resolution localization of NHE3 in adult and neonatal rat kidney. The American Journal of Physiology. 1997; **273**(2 Pt 2):F289-F299

[189] Amemiya M, Loffing J, Lotscher M, Kaissling B, Alpern RJ, Moe OW. Expression of NHE-3 in the apical membrane of rat renal proximal tubule and thick ascending limb. Kidney International. 1995;**48**(4):1206-1215

[190] Moe OW. Acute regulation of proximal tubule apical membrane Na/H exchanger NHE-3: Role of phosphorylation, protein trafficking, and regulatory factors. Journal of the American Society of Nephrology. 1999; **10**:2412-2425

[191] Schultheis PJ, Clarke LL, Meneton P, Miller ML, Soleimani M, Gawenis LR, et al. Renal and intestinal absorptive defects in mice lacking the NHE3 Na^+/H^+ exchanger. Nature Genetics. 1998; **19**(3):282-285

[192] Wu MS, Biemesderfer D, Giebisch G, Aronson PS. Role of NHE3 in mediating renal brush border Na^+/H^+ exchange. Adaptation to metabolic acidosis. The Journal of Biological Chemistry. 1996;**271**(51): 32749-32752

[193] Wang T, Yang CL, Abbiati T, Schultheis PJ, Shull GE, Giebisch G, et al. Mechanism of proximal tubule bicarbonate absorption in NHE3 null mice. The American Journal of Physiology. 1999;**277**(2 Pt 2):F298-F302

[194] Ledoussal C, Lorenz JN, Nieman ML, Soleimani M, Schultheis PJ, Shull GE. Renal salt wasting in mice lacking NHE3 Na^+/H^+ exchanger but not in mice lacking NHE2. American Journal of Physiology. Renal Physiology. 2001; **281**(4):F718-F727

[195] Woo AL, Noonan WT, Schultheis PJ, Neumann JC, Manning PA, Lorenz JN, et al. Renal function in NHE3-deficient mice with transgenic rescue of small intestinal absorptive defect. American Journal of Physiology. Renal Physiology. 2003;**284**(6):F1190-F1198

[196] Noonan WT, Woo AL, Nieman ML, Prasad V, Schultheis PJ, Shull GE, et al. Blood pressure maintenance in NHE3-deficient mice with transgenic expression of NHE3 in small intestine. American Journal of Physiology. Regulatory, Integrative and Comparative Physiology. 2005;**288**(3): R685-R691

[197] Li XC, Zhu D, Chen X, Zheng XW, Zhao C, Zhang JF, et al. Proximal tubule-specific deletion of the NHE3 (Na^+/H^+ exchanger 3) in the kidney attenuates angiotensin II-induced hypertension in mice. Hypertension. 2019 (in press)

Chapter 6

Scientific Evidences Supporting the Activation of the Renin-Angiotensin-Aldosterone System during Estral Cycle and Pregnancy in Mares

Katy Satué and Ana Muñoz

Abstract

In women and laboratory animals, local and circulating components of the renin-angiotensin-aldosterone system (RAAS) are related to specific reproductive functions that occur during the estrous cycle, such as folliculogenesis, ovulation, corpus luteum development, and steroidogenesis. Also, in pregnant females of these species, maternal cardiovascular and renal systems undergo intense modifications, with the aim of matching the increased energy requirements of the fetus and fetoplacental unit. Some of these changes can be the origin, and others the consequence of a new endocrine environment. The fetus and the placenta induce endocrine changes, with modifications in the protein, lipid, carbohydrate, and mineral metabolism, together with simultaneous cardiovascular changes derived from the uterine growth and its content. The participation of RAAS during this period is of vital importance to regulate these cardiovascular, hemodynamic, hematological, and metabolic adjustments imposed by pregnancy because they will have a direct influence on the correct development and viability of the fetus. In mares, our research team has been investigating the changes of RAAS in mares during the estral cycle and during pregnancy, and these results are presented in the current chapter, comparing with the data previously reported for women and laboratory animals.

Keywords: estrous cycle, mare, pregnancy, renin, angiotensin, aldosterone

1. Introduction

In nonpregnant females of various species, the components of the renin-angiotensin-aldosterone system (RAAS), i.e., prorenin, renin, angiotensin II (ANG-II), and aldosterone, are expressed in the tissues of reproductive organs, mainly the uterus and ovaries. These hormones have direct physiological relationships with specific reproductive functions, including folliculogenesis, oocyte maturation, ovulation, follicular atresia, corpus luteum development and luteolysis, steroidogenesis, angiogenesis, and expression of certain vasoactive substances [1–8].

A great body of literature has confirmed an increase in plasma activity of renin (PRA) and plasma concentrations of ANG-II and aldosterone in women during the

IntechOpen

luteal compared to the follicular period of the estrous cycle [9–19]. The primary source of renin and aldosterone is progesterone (P4) secreted by the corpus luteum. The elevated P4 concentrations during the luteal period lead to increased renal plasma flow, glomerular filtration rate, and natriuresis. This natriuretic effect stimulates in a compensatory way the synthesis and release of renin, ANG-II, and aldosterone [10–12, 15, 16, 20–23]. However, these results do not agree with those described by Szmuilowicz et al. [19], who suggested that the synthesis of aldosterone might be independent of renin and ANG-II at the ovarian level. Another physiological event that happens during the estrous cycle of the women is the preovulatory increase of renin [14, 24], ANG-II [25], and aldosterone [10, 12, 25, 26].

Similar results to those described in women have been reported in laboratory animals [27–30]. A potential physiological mechanism related to these changes is the stimulatory effect exerted by estrogens (E2) on the synthesis of angiotensinogen [31]. Additional mechanisms could be the hemodynamic and renal blood flow variations, changes in Na concentrations at the macula densa in the kidney, alterations in local sympathetic activity, and release of corticotropin or adrenocorticotrophic hormone (ACTH). All of these mechanisms act as regulatory factors for aldosterone synthesis [17, 19, 32, 33].

During pregnancy, circulating and tissular RAASs interact closely in order to achieve a successful outcome of the pregnancy. The local RAASs involved in pregnancy are mainly located at the ovaries, uterus (both at the placenta and decidua), and intrarenal level. The nonrenal local RAAS systems have pivotal functions in the implantation and in the placentation as well as in the development of the uteroplacental and umbilical placental circulations. In addition, they contribute directly to the circulating RAAS of the pregnant female, determining in part the normal functionality of the cardiovascular and renal systems [34].

In this review, we summarize the state of the art of the current knowledge of the RAAS in reproductive mares. We provide comparative profiling of the hormones of this system with other species, mainly with women and laboratory animals. The results of our own investigations performed during the last 10 years have demonstrated that an activation of the RAAS happens around ovulation and during pregnancy in the reproductive mare, even though significant differences with the reports made for women and laboratory animals have been found.

2. Changes in the renin-angiotensin-aldosterone system during the estral cycle

2.1 A brief review of the estral cycle in the mare

An exhaustive review of the estral cycle of the mare and the neuroendocrine mechanisms associated has been published recently [35]. Briefly, the mean length of a mare's estral cycle is 21 ± 3 days. The cycle is divided into two physiological periods: the estrous or follicular period and the diestrus or luteal period. The duration of the follicular period is approximately 6 days, but it can take 4–10 days depending on the mare. The luteal period has a duration of 15 days, ranging between 12 and 18 days. Ovulation can occur at any time during the follicular period, although it usually occurs 24–48 h before the end of this period. During the estral cycle, physiological changes in mares include follicular development, selection of the dominant follicle, release of the oocyte from follicle, formation of the corpus luteum, and production of P4. These physiological events are controlled by gonadotropin-releasing hormone (GnRH) that stimulates the hypophysis to secrete follicle-stimulating hormone (FSH) and luteinizing hormone (LH). These hormones control the development of the follicle and stimulate the secretion of E2 from the follicle to induce ovulation and corpus luteum establishment and development [35, 36].

2.2 Changes in the renin-angiotensin-aldosterone system during the luteal period

Most researchers have documented that aldosterone concentrations rise between 2 and 4 times during the luteal period [11, 37]. Presumably, this increase mirrors the increase experienced by ANG-II concentrations. However, this supposition has not always been supporting, and as a consequence, later it was suggested that the changes in ANG-II might be attributed to other ways independent of the RAAS [19]. Although this hypothesis is currently the most defended in the literature, some authors would not be able to confirm it [32, 38].

A variety of studies have shown that the relationship between PRA and aldosterone from the middle to the end of the luteal period can be adjusted to a linear model [9, 10, 16, 22, 37, 39, 40]. The researches performed in women with luteal insufficiency have helped to clarify complex issues, such as the involvement of the sex hormones in the modifications of the different RAAS components during the different period of the estral cycle. In this sense, several early investigations compared the concentrations of angiotensinogen, aldosterone, gonadotropins, ovarian steroids, and PRA in women with ovulatory cycles and in women with luteal insufficiency. While in physiological estral cycles the peak of PRA was closely related to the increase in aldosterone, in women with anovulatory cycles, these temporal relationships between renin and aldosterone or PRA and aldosterone were not found [9, 10, 41]. Subsequently, it was hypothesized that once the corpus luteum reaches functional maturity, it acts as the primary source of renin and aldosterone. Thereby, there was an explanation for the direct relationship between P4 and other compounds derived from the corpus luteum and the RAAS in physiological estral cycles, but not in women with ovulatory failure, because these women show a significant decrease in P4 concentrations [18, 21, 42].

P4 seems to be directly related to the luteal synthesis of aldosterone [15, 18, 21, 42]. Certain mechanisms dependent on P4, such as increased plasma flow, renal glomerular filtration rate, and Na and Cl excretion, result in natriuresis [16]. The initial natriuresis induced by P4 stimulates in a compensatory way the secretion of renin, ANG-II, and aldosterone [9–11, 20, 37]. The experimental administration of P4 results in natriuresis, which is followed by peaks of aldosterone secretion, aldosterone urinary excretion (AUE), and Na retention [11, 43]. Although PRA, ANG-II, and aldosterone increase after the administration of P4, the rises do not occur simultaneously [43]. These results have been explained in base of the competence between P4 and aldosterone for the mineralocorticoid receptor [12, 44, 45]. In physiological cycles, endogenously secreted P4 exerts a mild anti-aldosterone effect, avoiding an excessive retention of Na and water during the luteal period [46].

The E2 and P4 peaks during the luteal period lead to increases of PRA and ANG-II 2–3 times and also increases of AUE [17, 19]. However, the increase of E2 that happens during the first half of the estral cycle does not result in substantially raised PRA and AUE. Perhaps these events are mediated by the expression of aldosterone receptor protein and angiotensinogen in the absence of changes in ANG-II [46–48]. The lack of correlation between aldosterone and E2, as well as the lack of stimulation of aldosterone production in in vitro cultures of glomerulosa cells, could be other findings that might exclude E2 from the synthesis of aldosterone during the luteal period [19].

However, there are additional mechanisms involved in the synthesis of aldosterone independently of the P4 concentrations, such as the uptake of proteins and Na in the diet [49, 50] or a peripheral vasodilation [16]. It is difficult to interpret these findings without considering the intake of Na, since this electrolyte determines primarily the production of aldosterone via RAAS [49]. However, many previous studies did not control or did not report Na concentrations, which is a limiting factor in the proper interpretation of the results. More recently, Szmuilowicz et al. [19] showed that circulating and urinary levels of aldosterone increase significantly during the luteal period in response to the infusion of ANG-II in women with high Na concentrations. In these situations, positive correlations between P4 and aldosterone have been found, without parallel changes in PRA and ANG-II. On the contrary, in women with low Na concentrations, these interrelationships do not occur, even though the restriction of Na in the diet is a powerful stimulus for the activation of RAAS [49]. In Na-restricted diets, PRA and aldosterone respond independently between each other, and therefore, the absence of modifications in PRA and ANG-II could rule out that the increase in aldosterone is modulated via RAAS activation during the luteal period.

Another mechanism that has been related to the synthesis of renin and aldosterone during the luteal period is the release of angiotensinogen and prorenin. However, the studies conducted in this field have provided contradictory results. The expression of angiotensinogen is regulated transcriptionally by E2 [31], and consequently, the administration of different ovarian stimulation treatments in women increases the concentrations of E2 and renin [15]. Therefore, the activation of the RAAS during the luteal period could be partially related to the increased synthesis of angiotensinogen [21] or be a direct consequence of the vasodilatory effects of E2 [51]. However, early studies have not revealed variations in angiotensinogen during the luteal period of physiological estral cycles [10, 52, 53]. The combined use of E2 and P4 during the luteal period in ovariectomized rats resulted in increases in PRA without significant simultaneous changes in angiotensinogen [54]. The lack of a significant relationship between PRA and angiotensinogen might rule out E2 as a cause of the modifications in renin concentrations, despite its involvement in the secretion of aldosterone [10, 53, 55].

Although the main source of circulating active renin is the kidney, the ovaries appear to be the source of the cyclic increase of prorenin during the estral cycle [14, 56]. In fact, the release of LH triggers an increase of three times the concentrations of prorenin, remaining elevated during the first half of the luteal period, even though the decrease in P4 at the end of the luteal period reduces simultaneously the prorenin levels [18, 21, 42, 56].

2.3 Changes in the renin-angiotensin-aldosterone system during the follicular period

Preovulatory increases of angiotensinogen, prorenin, PRA, renin, ANG-II, and aldosterone have been reported in women [11, 14, 25, 26] and laboratory animals [29]. According to other authors, the increase in aldosterone concentrations appears to be transient [11], or it is not constant in all cycles [37].

The origin of the preovulatory aldosterone peak is unknown, but it has been speculated that the increase in E2 could exert a stimulatory effect on angiotensinogen in the liver, and this, in turn, increases PRA and aldosterone [52, 53, 55]. Other researches, on the contrary, failed to find any relationship between angiotensinogen, PRA, and aldosterone in physiological cycles [11, 17, 19, 57], ruling out some type of significant influence of endogenous E2 on PRA and AUE during the periovulatory period [9, 10, 41]. Despite these contradictory results, the hypothesis that the increase in angiotensinogen is the cause of PRA and aldosterone peaks is still commonly accepted.

A transient peak of prorenin has been described simultaneously with the elevation of gonadotropins, mainly LH, toward the moment of the ovulation. It has been speculated that this prorenin peak will be responsible later for the increase of active renin [58]. However, prorenin and LH are not always related to active renin [14, 15], because the most striking effects of this precursor appear toward the middle of the luteal period [18, 21, 42].

Because ACTH is one the main regulators of aldosterone secretion, the pre-ovulatory peak of PRA might be related to the stimulatory effect of E2 on the adrenal gland and to the increase of the 17α-P4-enzyme, which also shows a preovulatory peak. This enzyme catalyzes the conversion of pregnenolone into 17-α-hydroxypregnenolone, which is a precursor of the sexual steroids [59]. However, although an experimental infusion of ACTH triggers an elevation in PRA [60], the absence of changes in cortisol (CORT) concentrations in physiological cycles could refute these hypotheses [17, 19, 32]. Other factors, including hemodynamic changes in renal blood flow, local sympathetic activity, and variation in Na concentrations at the macula densa level, have also been associated with the preovulatory peak of renin and aldosterone [33, 61].

In broodmares, Satué et al. [62] evaluated the changes in circulating RAAS components around ovulation, considering the 5 days before ovulation (from day 5 to day 1), the day of the ovulation (day 0), and the 5 days immediately after ovulation (from day 1 to day 5). A summary of the main endocrine changes, including those of the RAAS, is presented in **Figure 1**.

A progressive increase in plasma renin concentrations was found before ovulation, peaking at the day of the ovulation [62] (**Figure 2**). The increase in plasma renin before ovulation in other species, as has been previously attributed to the release of LH or to the stimulating effect of human chorionic gonadotropin (hCG) [24]. Another plausible explanation for this rise in plasma renin before ovulation was the increase in E2 concentrations, since a positive correlation between both variables has been found in mares (r = 0.744) [62] (**Figure 3**). These results in mares agree with the data provided by Sealy et al. [15] in women with ovarian stimulation, in whom significant simultaneous increases in renin and E2 were detected. Another explanation for the rise in renin concentrations in the mares could be a mild hypovolemia associated with the presence of interstitial fluid in

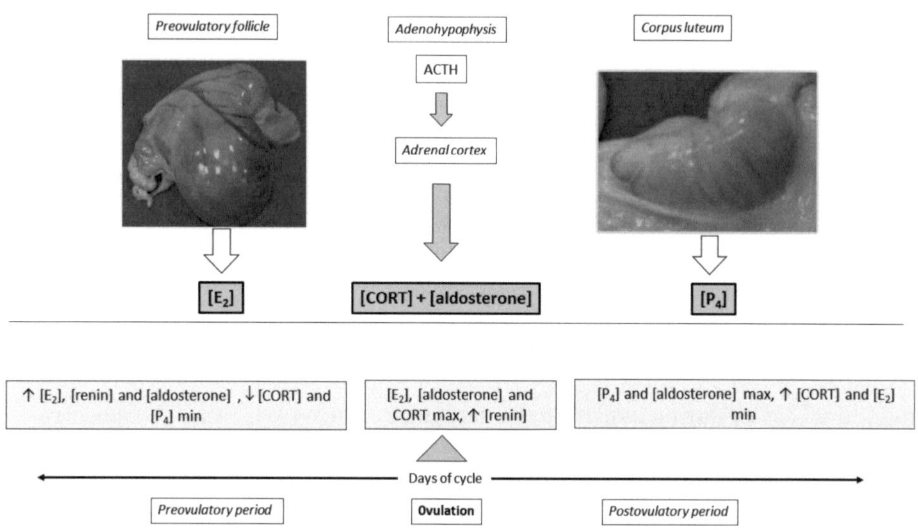

Figure 1.
Representative scheme of the hormonal changes that occur in the mare during the 5 days before ovulation (preovulatory period), the day of the ovulation, and the last first 5 days after ovulation (postovulatory period) (ACTH, adrenocorticotrophic hormone; E2, estrogens; CORT, cortisol; P4, progesterone).

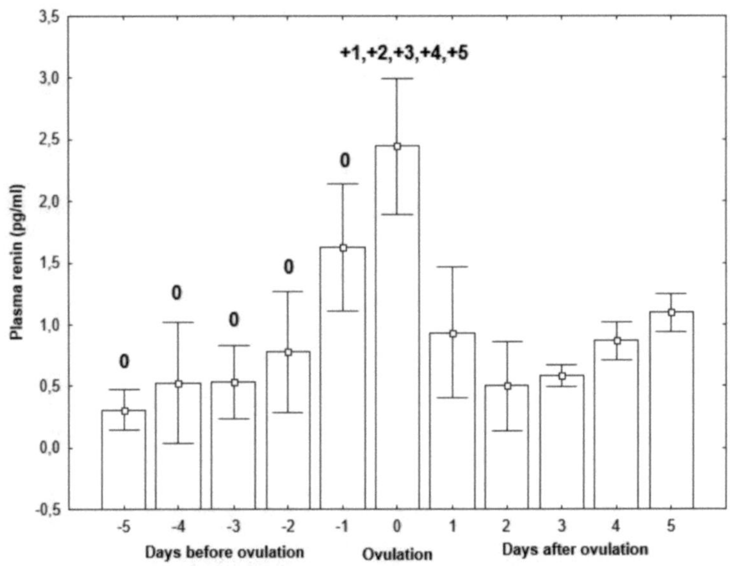

Figure 2.
Plasma renin concentrations during the 5 days before ovulation, the day of the ovulation, and the first 5 days after ovulation in mares (numbers indicate the days between which significant differences were found at $p < 0.05$*).*

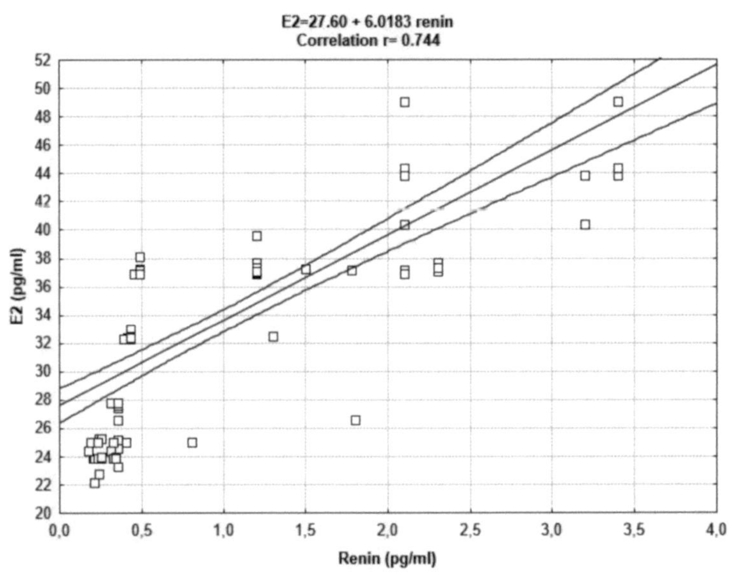

Figure 3.
Significant positive correlations between plasma estrogens (E2) and renin concentrations in mares before ovulation ($r = 0.744$*).*

abdominal cavity during ovulation. Even though changes in Na and Cl concentrations at the renal juxtaglomerular apparatus significantly influence the release of renin [63], the correlations between renin and these electrolytes appear to be low in mares before ovulation [62]. In the same way, Roussel et al. [64] described cyclic patterns of plasma Na and aldosterone concentrations in cow, but without significant correlations with the components of the RAAS.

After ovulation, a sharp decrease in plasma renin concentrations was detected in mares (**Figure 2**). Similar changes have been reported in women [17]. Ounis-Skali et al.

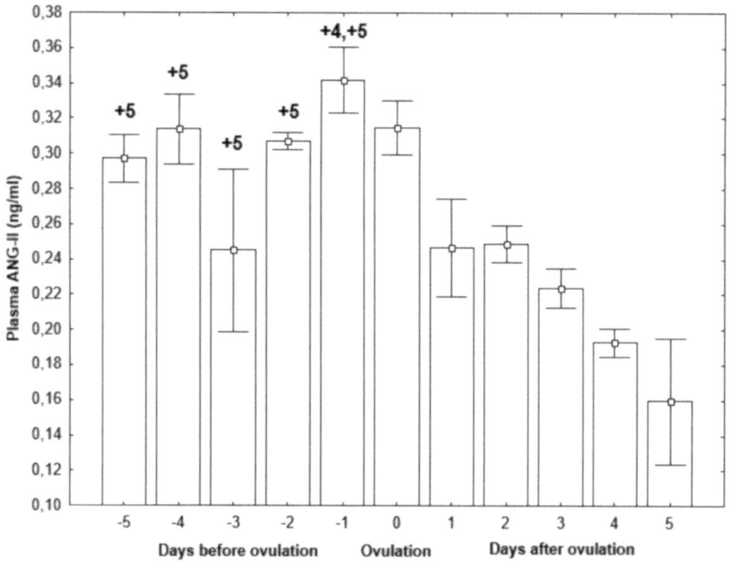

Figure 4.
Plasma angiotensin-II concentrations during the 5 days before ovulation, the day of the ovulation, and the first 5 days after ovulation in mares (numbers indicate the days between which significant differences were found at $p < 0.05$).

[18] attributed these results to the release of renin from the corpus luteum, reflecting a direct relationship between P4 and renin after ovulation, and similar hypotheses might be extrapolable to the mare. Two studies performed by our research team demonstrated low correlations between P4 and renin, suggesting other factors different from P4 might be considered in order to explain the non-significant trend toward an increase in renin concentrations during the first days after ovulation [65, 66].

Plasma ANG-II concentrations did not show significant variations before ovulation, but it experienced a sharp and progressive decrease after ovulation [62] (**Figure 4**). On the contrary, significant increases in ANG-II have been previously observed in sows [11] and in rats [29] before ovulation. ANG-II concentrations in women after ovulation were almost the double of the values found before ovulation [17–19]. This increase in women was attributed to the modulation of the flow in the ovarian blood vessels during oocyte maturation [15]. The reason why ANG-II concentrations decrease in the mares after ovulation, which is an opposite change to this described for women and laboratory animals, is not currently known.

Plasma aldosterone concentrations increased progressively from the fifth day before ovulation until the day of the ovulation. After this moment, aldosterone concentrations did not change and persisted to be significantly higher than before ovulation [62] (**Figure 5**). Because an increase in renin was also found the day of the ovulation in mares and ANG-II was higher during the 5 days before ovulation, it was speculated that the increased renin was the starting point to activate RAAS, resulting in increased aldosterone concentrations. This hypothesis was proven when the correlations between hormones before and after ovulation were assessed (**Figure 6**). A positive significant correlation ($r = 0.756$) was found between renin and aldosterone concentrations before ovulation. Surprisingly, ANG-II and aldosterone concentrations did not present a significant correlation before ovulation (**Figure 6**). These data might indicate a dissociation between ANG-II and aldosterone before ovulation in the mare. In this case, the synthesis of aldosterone might have occurred in addition via other pathways independent from the activation of the RAAS [62].

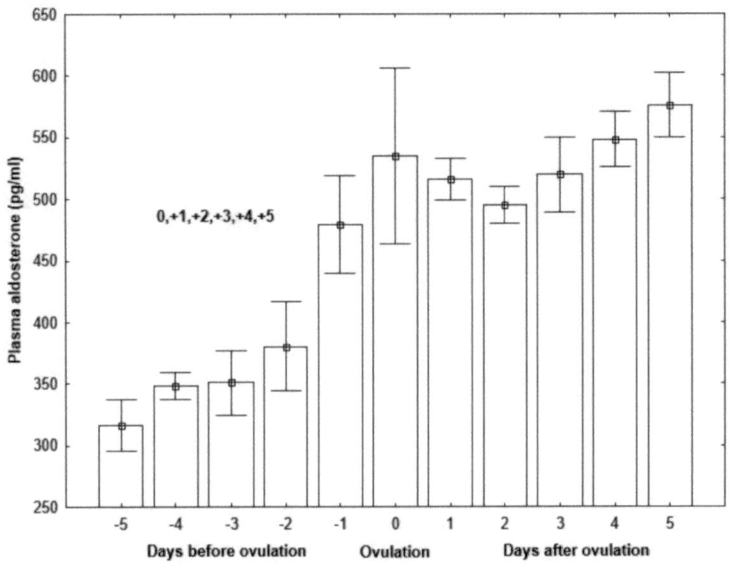

Figure 5.
Plasma aldosterone concentrations during the 5 days before ovulation, the day of the ovulation, and the first 5 days after ovulation in mares (numbers indicate the days between which significant differences were found at $p < 0.05$).

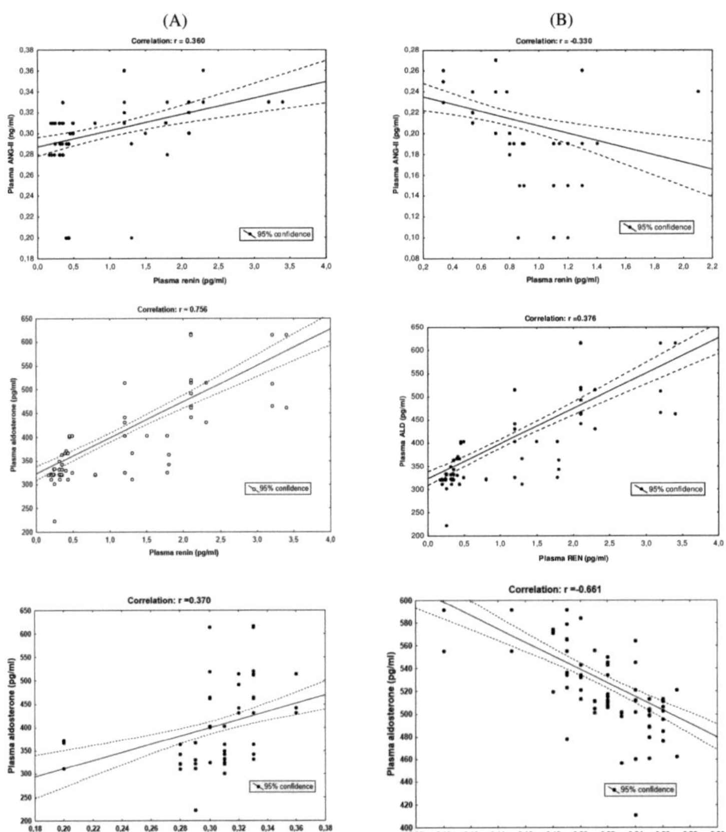

Figure 6.
Correlations between the three main components of the renin-angiotensin-aldosterone system in mares before (A) and after ovulation (B).

The relationships between the diameter of the predominant follicle and the hormones of the RAAS have also been investigated in mares [62] (**Figure 7**). A significant positive correlation was found between renin and aldosterone concentrations and the diameter of the follicle before ovulation in the mares (r = 0.742 and 0.804, respectively) [62] (**Figure 7**). In other species, RAAS regulates the development and growth of the antral follicles. Receptors AT1 and AT2 for ANG-II have been found in the granulosa cells of the antral follicles and in the primordial primary and secondary follicles of sows [67]. Administration of an ANG-II inhibitor in cows avoided the growth of the predominant follicle and also induced a reduction in the concentrations of E2 [68, 69]. In the same way, in women subjected to ovarian stimulation with

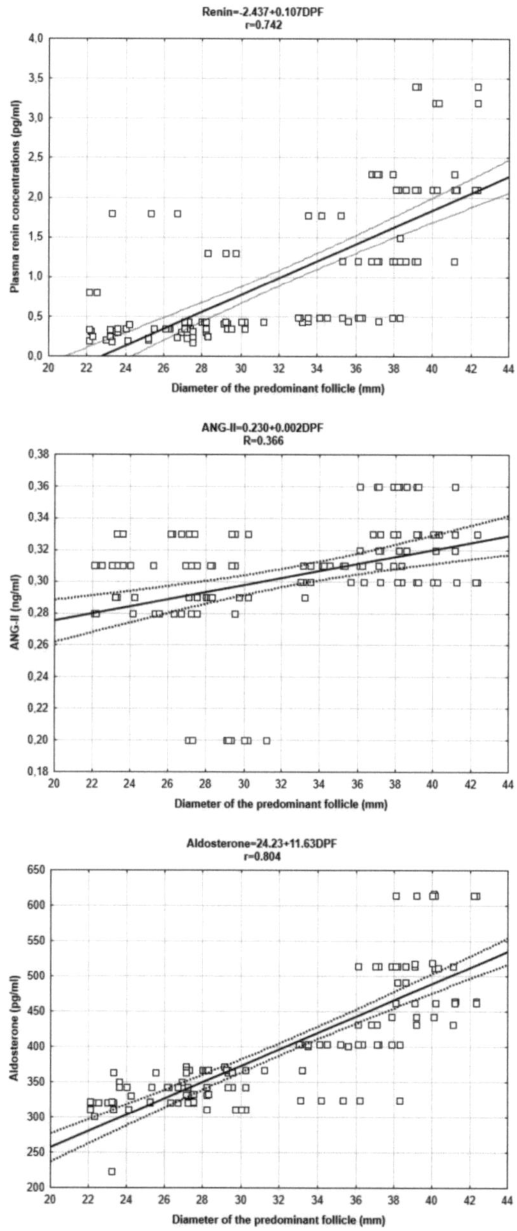

Figure 7.
Correlations between the diameter of the predominant follicle (DPF) and the concentrations of renin, angiotensin II, and aldosterone before ovulation in mares.

human menopausal gonadotropin (HMG), concentrations of ANG-II were positively correlated with E2 and the diameter of the predominant follicle [70]. In our research in mares, however, correlations between the diameter of the predominant follicle and ANG-II were very low (r = 0.366), despite the significant correlations between this diameter and the concentrations of renin and aldosterone [62] (**Figure 7**). Our results might suggest that circulating ANG-II concentrations do not exert a transcendental effect on the development of the ovarian follicle or, alternatively, some antagonist peptides of the ANG-II might be produced before ovulation in mares [65, 66].

3. Changes in the renin-angiotensin-aldosterone system during the pregnancy

Plasma concentrations of prorenin increase by 5–10 times during the first 10 weeks of pregnancy in women. This increase is simultaneous with the increase in hCG and after, both decreased until the moment of the delivery. Although prorenin is mainly synthesized in the renal juxtaglomerular cells, there are other extrarenal sources, such as the ovary, follicular fluid, and placenta [71]. The prorenin concentrations also increase when the LH concentrations peak at the moment of the ovulation [71, 72]. The combined administration of LH and FSH causes a 390% increase in plasma prorenin, while hCG increases above 1000% [13, 73]. An early study speculated that prorenin could act as a hormone with functions independent of the active renin [13]. Until now, to the authors' best knowledge, the prorenin concentration in plasma/serum has not been quantified in mares.

A substantial increase in the circulating concentrations of angiotensinogen has also been documented during the first weeks of pregnancy in women [74]. Angiotensinogen concentrations remain elevated throughout pregnancy, peaking at term. In sheep, a bimodal pattern of secretion of this hormone has been described, with a first peak at the beginning of the pregnancy, concomitant with the period of the maximum placental growth. The second peak occurs at term, suggesting an association between plasma angiotensinogen and fetal growth [75].

The increased synthesis of angiotensinogen has been attributed to the stimulating effect of E2 at hepatic level [16, 17, 76]. However, the relationships between angiotensinogen and E2 are not always positive [77]. In nonpregnant women undergoing estrogen treatments, angiotensinogen and ANG-II increased in blood, whereas renin concentrations tended to decrease because of the negative feedback exercised by angiotensinogen [76]. However, the angiotensinogen concentrations have not been measured in mares yet.

The total concentration of circulating renin, both its active and inactive forms, increases during pregnancy in women. However, the contribution of the physiologically inactive renin to the total renin is greater than that of the active renin [78]. During the first third period of the pregnancy, renin concentrations increased between 2 and 4 times above baseline [79, 80]. After it reaches a plateau, around the fifth month of pregnancy and after that, the concentrations remained unchanged until the end of the pregnancy [81], being the main reason involved in the rise of PRA during the first and second thirds of the pregnancy, respectively [11, 80, 82–84]. Toward the end of the pregnancy, the renin concentration decreases intensely [78]. Even though the increased renin release contributes greatly to the ANG-II synthesis, the renin concentrations could decrease because of a negative feedback mechanism [15, 85].

The synthesis of renin, in addition to renal juxtaglomerular cells, can also be carried out in other extrarenal tissues, as the uterus. This organ is the priority organ of renin synthesis during pregnancy, leading to the so-called hyperreninemia of the

DOI: http://dx.doi.org/10.5772/intechopen.88052

pregnancy [80, 86]. The renin at the level of the fetal membranes and the uterus is found predominantly in the form of prorenin, and although it is supposed that it can be released into the maternal circulation, the degree of contribution to the maternal circulating levels is unknown. Both active renin and inactive renin have been identified in the uterus, placenta, and myometrium in women and in some laboratory animals [87]. Despite this, amniotic fluid seems to be the main source of renin produced by the chorionic cells during pregnancy [74].

The placenta appears to be the major source for the conversion of angiotensin I into ANG-II, since the prorenin, active renin, angiotensin converting enzyme (ACE), and ANG-II have been identified in the chorion and in the placenta [88]. The physiological roles of these components have not been completely clarified, but it has been speculated that they might participate in the regulation of the blood flow [89–92].

The increase in renin concentrations during pregnancy has also been attributed to the loss of electrolytes as a consequence of the increase in the glomerular filtration rate as well as of the secretion of P4, due to the antagonic effects of the hormone on aldosterone, because P4 induces natriuresis [74, 76, 93]. On the other hand, the positive influence of E2 on the synthesis of angiotensinogen has also been considered a plausible reason for the renin peak during pregnancy [76]. Surprisingly, it is possible to find increased renin concentrations without a simultaneous increase of ANG-II concentrations. In addition, the sequestration of Na and water at the uterus might be another triggering factor for the release of renin during pregnancy, as documented in pregnant bitches [94].

In pregnant mares, Satué et al. [95] described a significant increase in the renin from the seventh month of pregnancy (**Figure 8**). In pregnant women, as stated before, the increase in renin concentrations occurs earlier than in the pregnant mares [79, 80, 84]. At present, it is unknown if the origin of these differences is the different animal species or it could be due to the influence of other physiological factors. In the mare, ANG-II concentrations did not vary significantly during pregnancy (**Figure 9**), despite the changes in renin concentrations (**Figure 8**). These data might suggest that there is a dissociation between renin and ANG-II during pregnancy in the mares. In pregnant women, it has been shown that the increase in

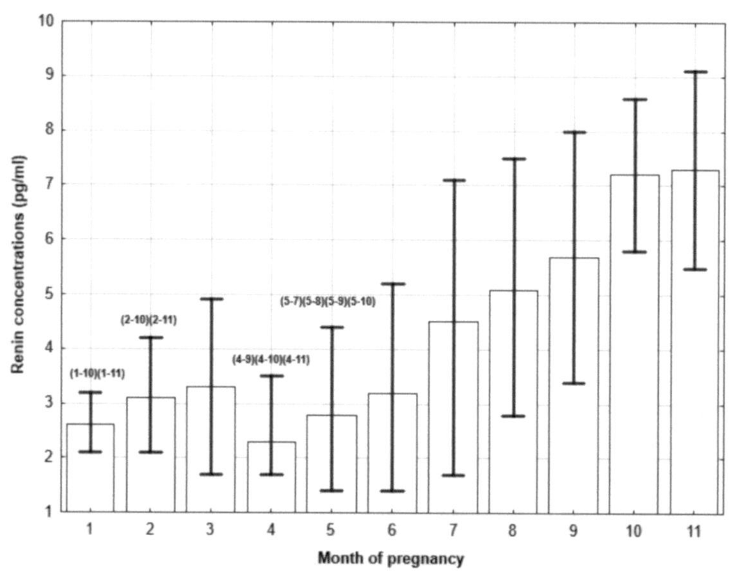

Figure 8.
Mean (maximum and minimum indicated in bars) concentrations of renin in mares during pregnancy (significant differences between months indicated with numbers at $p < 0.05$*).*

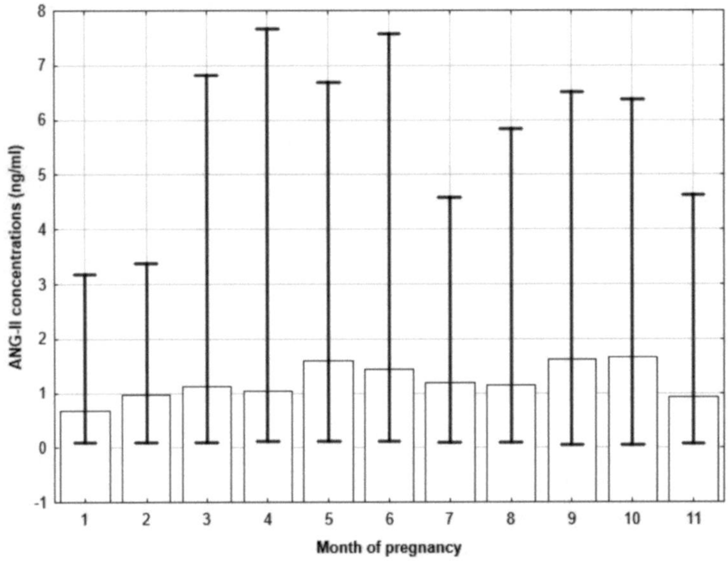

Figure 9.
Mean (maximum and minimum indicated in bars) concentrations of angiotensin II in mares during pregnancy.

ANG-II exerts a negative feedback effect on renin secretion [74, 85, 96]. In addition, it has been proposed that the peak of renin during pregnancy is in part associated with the release of estrogens, which act on the production of angiotensinogen in the liver [15, 76, 97]. Moreover, Satué et al. [98] found a dissociation between renin and E2, in agreement with previous results described for pregnant women [77, 99].

On the contrary to the data found for ANG-II in mares [95], most of the investigations performed in women and laboratory animals showed that during pregnancy, circulating concentrations of ANG-II increased up to two and even three times from nonpregnant values [74, 78, 100, 101]. This increase in women appears from the beginning of the pregnancy, peaking at the seventh month and persisting to be elevated until the moment of the delivery [102].

Pregnant mares showed a marked increase in circulating aldosterone concentrations during pregnancy, with mean values that exceed almost 12 times the physiological reference range for healthy adult horses [103] (**Figure 10**). Similar results have been reported in pregnant women [74, 84, 104], bitches [105–107], and some species of laboratory animals [108]. In pregnant bitches, Robb et al. [94] and in rats, Brochu et al. [106] attributed the increase in aldosterone concentrations to a higher activity of the enzyme aldosterone-synthase and to an increased synthesis of mRNA in the cells of the zona glomerulosa, these cells being the only production site, without fetal and/or placental participation, on the contrary of what appears to happen in women [78].

In mares, our team has found lower aldosterone concentrations during the second month of pregnancy, compared to the fifth, sixth, and seventh months of pregnancy (**Figure 10**). The highest values were observed in the fifth month of pregnancy. In the eighth, ninth, and tenth months of pregnancy, aldosterone concentrations were significantly lower than the mean values found in the fifth month. In pregnant women, it has been speculated that the increased release of aldosterone would occur in an attempt to conserve Na and water to favor the expansion of the fetoplacental unit. In this way, the adequate supply of nutrients to the fetus, the maintenance of the ideal oxygen tension for fetal development, as well as the homeostasis and appropriate blood pressure between the mother and the fetus would be guaranteed [108, 109].

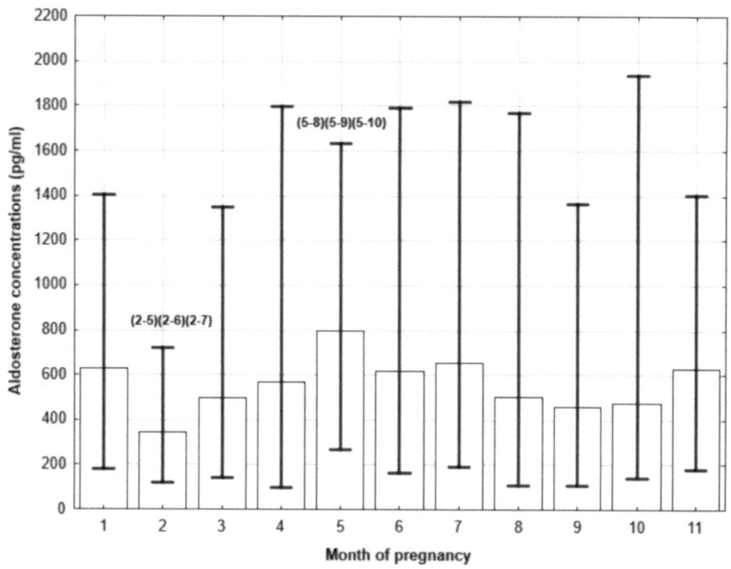

Figure 10.
Mean (maximum and minimum indicated in bars) concentrations of aldosterone in mares during pregnancy (significant differences between months indicated with numbers at $p < 0.05$).

Although the evolution of aldosterone concentrations during pregnancy is similar in women, laboratory animals, and mares, there are temporary differences in the moment in which the peaks of concentrations are reached. In women and laboratory animals, the maximum increase in aldosterone appears in the second third of pregnancy, followed by a decrease in the third period of pregnancy. In the mare, on the contrary, as shown in **Figure 10**, the maximum concentrations were found in the fifth month.

The increase in aldosterone during pregnancy in women has been attributed, firstly, to the increase in the sensitivity of the cells of the adrenal gland to the increased synthesis and release of renin [74, 110] and, secondly, to the increase in ANG-II concentrations, a finding that has been associated with an increase in placental lactogen [74]. In women, it is known that placental lactogen plays a regulatory role in metabolic homeostasis, triggering an increase in ANG-II receptors during the second period of the pregnancy [111, 112]. It has also been speculated in women that the increase in the concentrations of aldosterone during pregnancy could be a decisive physiological event in order to prevent the massive natriuresis that could arise from an enhanced glomerular filtration rate. In these cases, aldosterone shows a physiological action opposite to the natriuretic effect of P4 at the level of the distal convoluted tubule, avoiding excessive loss of Na and allowing its gradual accumulation in the fetoplacental and in the extracellular maternal fluids [113]. However, the antagonism between both mineralocorticoids, i.e., aldosterone and P4, during pregnancy does not seem to be constant [114, 115].

4. Conclusions

Our results obtained in healthy reproductive mares demonstrated that there is an activation of the RAAS around the time of ovulation, as indicated by the increased plasma renin and aldosterone concentrations. After ovulation, the rapid decrease in plasma renin and ANG-II concentrations might indicate a modulation of the previously activated RAAS, even though plasma aldosterone concentrations increased during this period, suggesting a dissociation of the components of this system.

In the pregnant mare, there is also an increased activity of various components of the RAAS. Pregnant mares show a progressive increase in circulating renin, with fluctuating changes in aldosterone, peaking at fifth month of pregnancy and without significant changes in ANG-II. The reason for the apparent dissociation between the three main hormones of the RAAS at determined moments of the pregnancy remains unclear. The data obtained in pregnant mares regarding the RAAS seems to indicate that renin is a more intense stimulus for the increased synthesis of aldosterone than ANG-II.

Acknowledgements

The authors would like to express their gratitude to the owners who have allowed, over a period of more than 10 years, the participation of their mares in the studies carried out by our research team.

Conflict of interest

The authors declare that they have no conflicts of interest.

Author details

Katy Satué[1] and Ana Muñoz[2]*

1 University Cardenal Herrera-CEU, Valencia, Spain

2 University of Córdoba, Spain

*Address all correspondence to: ksatue@uch.ceu.es

References

[1] Schauser KH, Nielsen AH, Dantzer V, Poulsen K. Angiotensin-converting enzyme activity in the bovine uteroplacental unit changes in relation to the cycle and pregnancy. Placenta. 2001;**22**:852-862. DOI: 10.1053/plac.2001.0740

[2] Brunswing-Spickenheier B, Mukhopadhyay AK. Local regulatory factors in regulation of ovarian function: Role of prorenin-renin-angiotensin-system. Indian Journal of Experimental Biology. 2003;**41**(7):669-681

[3] Ferreira R, Oliveira J, Fernandes R, Moraes J, Gonçalves P. The role of angiotensin II in the early stages of bovine ovulation. Reproduction. 2007;**134**:713-719. DOI: 10.1530/REP-07-0239

[4] Hurley ME, Critelli M, Johnson A, DenHaese D, Jones C, Olivieri MP. Renin immunofluorescence in the metestrus rat uterus. The FASEB Journal. 2009;**23**(1):637.9-637.9. DOI: 10.1096/fasebj.23.1_supplement.637.9

[5] Pereira VM, Reis FM, Santos RA, Cassali GD, Santos SH, Honorato-Sampaio K, et al. Gonadotropin stimulation increases the expression of angiotensin (1-7) and MAS receptor in the rat ovary. Reproductive Sciences. 2009;**16**(12):1165-1174. DOI: 10.1177/1933719109343309

[6] Pountain SJ, Pipkin FB, Hunter MG. The ontogeny of components of the renin-angiotensin system in the porcine fetal ovary. Animal Reproduction Science. 2010;**117**(1-2):119-126. DOI: 10.1016/j.anireprosci.2009.03.006

[7] Herr D, Duncan WC, Hack G, Konrad R, Kreienberg R, Wulff C. Regulated expression of the renin-angiotensin-system in human granulosa lutein cells: Angiotensin II increases VEGF expression but its synthesis is reduced by hCG. Archives of Gynecology and Obstetrics. 2010;**281**(3):409-416

[8] Sneeringer R, Penzias AS, Barrett B, Usheva A. High levels of mineralocorticoids in preovulatory follicular fluid could contribute to oocyte development. Fertility and Sterility. 2011;**95**(1):182-187. DOI: 10.1016/j.fertnstert.2010.06.028

[9] Sundsfjord JA, Aakvaag A. Plasma angiotensin II and aldosterone excretion during the menstrual cycle. Acta Endocrinologica. 1970;**64**:452-458

[10] Sundsfjord JA, Aakvaag A. Plasma rennin activity, plasma rennin substrate and urinary aldosterone excretion in the menstrual cycle in relation to the concentration of progesterone and oestrogens in the plasma. Acta Endocrinologica. 1972;**71**:519-529

[11] Sundsfjord JA, Aakvaag A. Variations in plasma aldosterone and plasma rennin activity throughout the menstrual cycle, with especial reference to the pre-ovulatory period. Acta Endocrinologica. 1973;**73**:499-508

[12] Oelkers W, Schoneshofer M, Blumel A. Effects of progesterone and four synthetic progestogens on sodium balance and the renin-aldosterone system in man. The Journal of Clinical Endocrinology and Metabolism. 1974;**39**(5):882-890. DOI: 10.1210/jcem-39-5-882

[13] Sealey JE, Atlas SA, Glorioso N, Manapat H, Laragh JH. Cyclical secretion of prorenin during the menstrual cycle: Synchronization with luteinizing hormone and progesterone. PNAS. 1985;**82**:8705-8709. DOI: 10.1073/pnas.82.24.8705

[14] Sealey JE, Cholst I, Glorioso N, Troffa C, Weintraub ID, James G, et al. Sequential changes in plasma luteinizing

hormone and plasma prorenin during the menstrual cycle. The Journal of Clinical Endocrinology and Metabolism. 1987;**65**(1):1-5. DOI: 10.1210/jcem-65-1-1

[15] Sealey JE, Itskovitz-Eldor J, Rubattu S, James G, August P, Thaler I, et al. Estradiol and progesterone related increases in the renin aldosterone system: Studies during ovarian stimulation and early pregnancy. The Journal of Clinical Endocrinology and Metabolism. 1994;**79**(1):258-264. DOI: 10.1210/jcem.79.1.8027239

[16] Chapman AB, Zamudio S, Woodmansee W, Merouani A, Osorio F, Johnson A, et al. Systemic and renal hemodynamic changes in the luteal phase of the menstrual cycle mimic early pregnancy. American Journal of Physiology-Renal Physiology. 1997;**273**(5):777-782

[17] Chidambaram M, Duncan J, Lai V, Cattran D, Floras J, Scholey J, et al. Variation in the rennin angiotensin system throughout the normal menstrual cycle. Journal of the American Society of Nephrology. 2002;**13**:446-452

[18] Ounis-Skali N, Mitchell GF, Solomon CG, Solomon SD, Seely EW. Changes in central arterial pressure waveforms during the normal menstrual cycle. Journal of Investigative Medicine. 2006;**54**(6):321-326. DOI: 10.2310/6650.2006.05055

[19] Szmuilowicz E, Adler G, Williams J, Green D, Yao T, Hopkins P, et al. Relationship between aldosterone and progesterone in human menstrual cycle. The Journal of Clinical Endocrinology and Metabolism. 2006;**91**(10):3981-3987. DOI: 10.1210/jc.2006-1154

[20] Oelkers W, Berger V, Bolik A, Bahr V, Hazard B, Beier S, et al. Dihydrospirorenone, a new progestogen with antimineralocorticoid activity: Effects on ovulation, electrolyte excretion, and the renin-aldosterone system in normal women. The Journal of Clinical Endocrinology and Metabolism. 1991;**73**(4):837-842. DOI: 10.1210/jcem-73-4-837

[21] Oelkers W. Effects of estrogens and progestogens on the rennin-aldosterone system and blood pressure. Steroids. 1996;**61**(4):166-171

[22] Hirshoren N, Tzoran I, Makrienko I, Edoute Y, Plawner MM, Itskovitz-Eldor J, et al. Menstrual cycle effects on the neurohumoral and autonomic nervous systems regulating the cardiovascular system. The Journal of Clinical Endocrinology and Metabolism. 2002;**87**(4):1569-1575. DOI: 10.1210/jcem.87.4.8406

[23] Nakamura T, Okamura N, Yagi M, Omatsu H, Yamamori M, Kuwahara A, et al. Effects of ABCB1 3435C>T genotype on serum levels of cortisol and aldosterone in women with normal menstrual cycles. Genetics and Molecular Research. 2009;**8**(2):397-403

[24] Loret de Mola JR, Goldfarb JM, Hecht BR, Babbo CJ, Friedlander MA. Gonadotropins induce higher active renin levels in the follicular fluid of normal and hyperstimulated cycles. Gynecological Endocrinology. 1999;**13**:155-160

[25] Weir RJ, Doig A, Fraser R, Morton JJ, Parboosing J, Robertson JI, et al. Studies of the renin-angiotensin-aldosterone system, cortisol, DOC, and ADH in normal and hypertensive pregnancy. Perspectives in Nephrology and Hypertension. 1976;**5**:251-261

[26] Frölich JC, Hollifield JW, Dormois JC, Frölich BL, Seyberth H, Michelakis AM, et al. Suppression of plasma renin activity by indomethacin in man. Circulation Research. 1976;**39**:447-452. DOI: 10.1161/01.RES.39.3.447

[27] Howard RB, Pucell AG, Bumpus FM, Husain A. Rat ovarian renin: Characterization and changes during the estrous cycle. Endocrinology. 1988;**123**(5):2331-2340. DOI: 10.1210/endo-123-5-2331

[28] De Vito E, Cabrera RR, Guardia DC. Changes during lactation in the renin-like enzyme concentration in rat luteal tissue. Acta Endocrinologica. 1988;**116**(4):526-530

[29] De Vito E, Guardia DC, Cabrera RR. Cyclical changes in plasma renin during the oestrous cycle in rats: Synchronized effect of oestrogen and progesterone. The Journal of Endocrinology. 1989;**121**(2):261-267

[30] Harewood W, Gillin A, Mohamed S, Willis N, Lazzaro V, Duggin G, et al. Cyclical changes in the rennin-angiotensin-aldosterone system during the menstrual cycle of the baboon (*Papio hamadryas*). Journal of Medical Primatology. 1996;**25**(4):267-271

[31] Gordon MS, Chin WW, Shupnik MA. Regulation of angiotensinogen gene expression by estrogen. Journal of Hypertension. 1992;**10**:361-366

[32] Pechere-Bertschi A, Maillard M, Stalder H, Brunner H, Burnier M. Renal segmental tubular response to salt during the normal menstrual cycle. Kidney International. 2002;**61**:425-431. DOI: 10.1046/j.1523-1755.2002.00158.x

[33] Sealey JE, Laragh JH. Plasma renin activity enzyme-kinetic assay: Protection of angiotensin I from bacterial degradation. Clinical Chemistry. 2011;**57**(3):529-530. DOI: 10.1373/clinchem.2010.156596

[34] Lumbers ER, Pringle KG. Roles of the circulating renin-angiotensin-aldosterone system in human pregnancy. American Journal of Physiology. Regulatory, Integrative and Comparative Physiology. 2014;**306**(2):91-101. DOI: 10.1152/ajpregu.00034.2013

[35] Satué K, Gardón JC. A review of the estrous cycle and the neuroendocrine mechanisms in the mare. Journal of Steroids and Hormonal Science. 2013;**4**:115. DOI: 10.4172/2157-7536.1000115

[36] McKinnon AO, Squires EL, Vaala WE, Dickson DV. Equine Reproduction. 2nd ed. Ames, IA: Wiley Blackwell; 2011. pp. 3288-3299. ISBN: 978-0-813-81971-6

[37] Katz FH, Romfh P. Plasma aldosterone and renin activity during the menstrual cycle. The Journal of Clinical Endocrinology and Metabolism. 1972;**34**:819-822. DOI: 10.1210/jcem-34-5-819

[38] Parry BL, Javeed S, Laughlin GA, Hauger R, Clopton P. Cortisol circadian rhythms during the menstrual cycle and with sleep deprivation in premenstrual dysphoric disorder and normal control subjects. Biological Psychiatry. 2000;**48**(9):920-931

[39] Michelakis AM, Yoshida H, Dormois JC. Plasma rennin activity and plasma aldosterone during the normal menstrual cycle. American Journal of Obstetrics and Gynecology. 1975;**123**(7):724-726

[40] Stachenfeld NS, DiPietro L, Kokoszka CA, Silva C, Keefe DL, Nadel ER. Physiological variability of fluid-regulation hormones in young women. Journal of Applied Physiology. 1999;**86**:1092-1096. DOI: 10.1152/jappl.1999.86.3.1092

[41] Sundsfjord JA. Plasma renin activity and aldosterone excretion during prolonged progesterone administration. Acta Endocrinologica. 1971;**67**(3):483-490

[42] Garcia MJ, Martinez-Martos JM, Mayas MD, Carrera MP, De la Chica S,

Cortes P, et al. Hormonal status modifies renin-angiotensin system regulating aminopeptidases and vasopressin-degrading activity in the hypothalamus-pituitary-adrenal axis of female mice. Medicinal Chemistry. 2008;**4**:336-347

[43] Stachenfeld NS, Taylor HS. Oestrogen effects on urine concentrating response in young women. The Journal of Physiology. 2003;**1**(3):869-880. DOI: 10.1113/jphysiol.2003.046920

[44] Myles K, Funder JW. Progesterone binding to mineralocorticoid receptors: In vitro and in vivo studies. The American Journal of Physiology. 1996;**270**(4):601-607. DOI: 10.1152/ajpendo.1996.270.4.E601

[45] Quinkler M, Meyer B, Bumke-Vogt C, Grossmann C, Gruber U, Oelkers W, et al. Agonistic and antagonistic properties of progesterone metabolites at the human mineralocorticoid receptor. European Journal of Endocrinology. 2002;**146**(6):789-799

[46] Keam S, Wagstraff A. Ethinylestradiol/drospirenone. A review of its use as an oral contraceptive. Treatments in Endocrinology. 2002;**2**(1):1-23

[47] Nowaczynski W, Murakami T, Richardson K, Genest J. Increased aldosterone plasma protein binding in women on combined oral contraceptives throughout the menstrual cycle. Journal of Clinical Endocrinology and Metabolism. 1978;**47**(1):193-199. DOI: 10.1210/jcem-47-1-193

[48] Williamson PM, Buddle ML, Brown MA, Whitworth JA. Ambulatory blood pressure monitoring (ABPM) in the normal menstrual cycle and in women using oral contraceptives. Comparison with conventional blood pressure measurement. American Journal of Hypertension. 1996;**9**(1):953-958. DOI: 10.1016/0895-7061(96)00150-1

[49] Adler GK, Moore TJ, Hollenberg NK, Williams GH. Changes in adrenal responsiveness and potassium balance with shifts in sodium intake. Endocrine Research. 1987;**13**(4):419-445

[50] Woods LL. Mechanisms of renal vasodilation after protein feeding: Role of the renin-angiotensin system. The American Journal of Physiology. 1993;**264**(3):601-609. DOI: 10.1152/ajpregu.1993.264.3.R601

[51] Mccaffrey TA, Czaja JA. Diverse effects of estradiol-17 beta: Concurrent suppression of appetite, blood pressure and vascular reactivity in conscious, unrestrained animals. Physiology & Behavior. 1989;**45**(3):649-657

[52] Skinner SL, Lumbers ER, Symonds EM. Alteration by oral contraceptives of normal menstrual changes in plasma renin activity, concentration and substrate. Clinical Science. 1969;**36**(1):67-76

[53] Katz FH, Romfh P, Smith P. Diurnal variation of plasma aldosterone, cortisol and renin activity in supine man. Journal of Endocrinology and Metabolism. 1975;**40**:125-134. DOI: 10.1210/jcem-40-1-125

[54] Ojeda N, Grigore D, Robertson E, Alexander B. Estrogen protects against increased blood pressure in postpubertal female growth restricted offspring. Hypertension. 2007;**50**:679-685. DOI: 10.1161/HYPERTENSIONAHA.107.091785

[55] Katz FH, Kappas A. The effects of estradiol and estriol on plasma levels of cortisol and thyroid hormone-binding globulins and on aldosterone and cortisol secretion rates in man. The Journal of Clinical Investigation. 1967;**46**(11):1768-1777. DOI: 10.1172/JCI105667

[56] Sealey JE, Parra D, Rosenstein R, Laragh JH. Effective plasma renin activity: A derived measure for assessing residual plasma renin activity in patients taking angiotensin-converting enzyme inhibitors or angiotensin receptor blockers. Hypertension. 2010;**55**(3):16-17. DOI: 10.1161/HYPERTENSIONAHA.109.148023

[57] Kaulhausen H, Leyendecker G, Benker G, Breuer H. The relationship of the renin-angiotensin-aldosterone system to plasma gonadotropin, prolactin, and ovarian steroid patterns during the menstrual cycle. Archiv für Gynäkologie. 1978;**225**:179-200

[58] Bouhnnik J, Feherentz JA, Galen FX, Seyer R, Evin G, Castro B, et al. Immunologic identification of both plasma and human renal inactive renin as prorenin. The Journal of Clinical Endocrinology and Metabolism. 1985;**60**(2):399-401. DOI: 10.1210/jcem-60-2-399

[59] Strott CA, West CD, Nakagawa K, Kondo T, Tyler FH. Plasma 11-deoxycorticosteroid and ACTH response to metyrapone (plasma metyrapone test). The Journal of Clinical Endocrinology and Metabolism. 1969;**29**(1):6-11. DOI: 10.1210/jcem-29-1-6

[60] Kem DC, Gomez-Sanchez C, Kramer NJ, Holland OB, Higgins JR. Plasma aldosterone and renin activity response to ACTH infusion in dexamethasone-suppressed normal and sodium-depleted man. The Journal of Clinical Endocrinology and Metabolism. 1975;**40**(1):116-124. DOI: 10.1210/jcem-40-1-116

[61] Yang S, Bae L, Zhang L. Estrogen increases eNOS and NOx release in human coronary artery endothelium. Journal of Cardiovascular Pharmacology. 2000;**36**(2):242-247

[62] Satué K, Montesinos P, Muñoz A. Activation of the renin-angiotensin-aldosterone system in mares around ovulation. Czech Journal of Animal Science. 2017;**62**(6):259-267. DOI: 10.17221/32/2016-CJAS

[63] Muñoz A, Riber C, Trigo P, Castejón-Riber C, Castejón FM. Dehydration, electrolyte imbalances and renin-angiotensin-aldosterone-vasopressin axis in successful and unsuccessful endurance horses. Equine Veterinary Journal. 2010;**38**:83-90. DOI: 10.1111/j.2042-3306.2010.00211.x

[64] Roussel JD, Clement TJ, Aranas TJ. Changes of aldosterone in blood serum of dairy cattle during estrous cycle. Journal of Dairy Science. 1983;**66**:1734-1747. DOI: 10.3168/jds.S0022-0302(83)81999-7

[65] Satué K, Montesinos P, Gardón JC. Association between aldosterone and cortisol levels during ovulatory period in Spanish purebred mares. Reproduction, Fertility, and Development. 2014;**26**:146. DOI: 10.1071/RDv26n1Ab65

[66] Satué K, Gardon JC, Montesinos P. Relationship between oestradiol-17beta and renin concentrations during preovulatory period in Spanish purebred mares. Reproduction in Domestic Animals. 2012;**47**(3):111. DOI: 10.1111/j.1439-0531.2012.02044.x

[67] Shuttleworth G, Hunter MG, Robinson G, Brouhton Pipkin F. I mmunocytochemical localization of angiotensin II receptor subtypes 1 and 2 in the porcine fetal, prepubertal and postpubertal ovary. Journal of Anatomy. 2002;**201**:267-274. DOI: 10.1046/j.1469-7580.2002.00091.x

[68] Ferreira R, Gasperin B, Rovani M, Santos J, Barreta M, Bohrer R, et al. Angiotensin II signaling promotes follicle growth and dominance in cattle. Endocrinology. 2011;**152**(12):4957-4965. DOI: 10.1210/en.2011-1146

[69] Ferreira R, Gasperin B, Santos J, Rovani M, Santos RA, Gutierrez K, et al. Angiotensin II profile and mRNA encoding RAS proteins during bovine follicular wave. Journal of the Renin-Angiotensin-Aldosterone System. 2011;**12**(4):475-482. DOI: 10.1177/1470320311403786

[70] Jarry H, Meyer B, Holzapfel G, Hinney B, Kuhn W, Wuttke W. Angiotensin II/III and substance P in human follicular fluid obtained during IVF: Relation of the peptide content with follicular size. Acta Endocrinologica. 1988;**119**:277-282

[71] Vrtacnik-Bokal E, Meden-Vrtovec H, Verdenik I. Uterine arterial blood flow and the substances of ovarian renin-angiotensin system in women with polycystic ovaries undergoing in vitro fertilization. European Journal of Obstetrics & Gynecology and Reproductive Biology. 2006;**124**(1):77-81

[72] Vrtacnik-Bokal EV, Vrotec HM, Virant Klun I, Verdenik I. Prolonged HCG action affects angiogenic substances and improves follicular maturation, oocyte quality and fertilization competence in patients with polycystic ovary syndrome. Human Reproduction. 2005;**20**:1562-1568. DOI: 10.1093/humrep/deh789

[73] Itskovitz J, Sealey JE, Glorioso N, Rosenwaks Z. Plasma prorenin response to human chorionic gonadotropin in ovarian-hyperstimulated women: Correlation with the number of ovarian follicles and steroid hormone concentrations. Proceedings of the National Academy of Sciences of the United States of America. 1987;**84**(20):7285-7289. DOI: 10.1073/pnas.84.20.7285

[74] Alhenc-Gelas F, Tache A, Saint-Andre JP, Milliez J, Sureau C, Corvol P, et al. The renin-angiotensin system in pregnancy and parturition. Advances in Nephrology from the Necker Hospital. 1986;**15**:25-33

[75] Dandrea J, Cooper S, Ramsay M, Keller-Woods M, Broughton Pipkin F, Symonds M, et al. The effects of pregnancy and maternal nutrition on the maternal renin angiotensin system in sheep. Experimental Physiology. 2002;**87**(3):353-359

[76] Oelkers W. The renin–aldosterone system and drospirenone. Gynecological Endocrinology. 2002;**16**:83-87

[77] Weinberger MH, Kramer NJ, Grim CE, Petersen LP. The effect of posture and saline loading on plasma renin activity and aldosterone concentration in pregnancy, non-pregnant and oestrogen treated women. The Journal of Clinical Endocrinology and Metabolism. 1977;**44**:69-77. DOI: 10.1210/jcem-44-1-69

[78] Carr B, Gant N. The endocrinology of pregnancy induced hypertension. Clinics in Perinatology. 1983;**10**:737-761

[79] Tapia HR, Johnson CE, Strong CG. Renin angiotensin system in normal and in hypertensive disease of pregnancy. Lancet. 1972;**2**:847-850. DOI: 10.1016/S0140-6736(72)92211-8

[80] Skinner SL. The renin system in fertility and normal human pregnancy. In: Robertson JIS, Nicolls MG, editors. The Renin Angiotensin System. London: Gower Medical Publishing; 1993. pp. 1-16

[81] Langer B, Grima M, Coquard C, Bader AM, Schlaeder G, Imbs JL. Plasma active renin, angiotensin I, and angiotensin II during pregnancy and in preeclampsia. Journal of Obstetrics and Gynaecology. 1998;**91**:196-202

[82] Weinberger MH, Kramer NJ, Petersen LP, Cleary RE, Young PC. Sequential changes in the renin-angiotensin-aldosterone systems and

plasma progesterone concentration in normal and abnormal human pregnancy. Perspectives in Nephrology and Hypertension. 1976;**5**:263-269

[83] Sullivan C, Matrin J. Sodium and pregnancy. Clinical Obstetrics and Gynecology. 1994;**37**:558-573

[84] Bentley-Lewis MD, Graves SW, Seely EW. The renin aldosterone response to stimulation and suppression during normal pregnancy. Hypertension in Pregnancy. 2005;**24**:1-16. DOI: 10.1081/PRG-45765

[85] Broughton Pipkin F, Morrison R, O'Brien PMS. The effect of prostaglandin E1 upon the pressor and hormonal response to exogenous angiotensin II in human pregnancy. Clinical Science. 1987;**72**:351-357

[86] Lindheimer MD, Katz AI. Renal physiology and disease in pregnancy. In: Seldin DW, Giebisch G, editors. The Kidney: Physiology and Pathophysiology. New York: Raven; 1992. pp. 3371-3432

[87] Skinner SL, Lumbers ER, Symonds EM. Renin concentration in human fetal and maternal tissues. American Journal of Obstetrics and Gynecology. 1968;**101**:529-533

[88] Kalenga MK, Thomas K, de Gasparo M, De Hertogh R. Determination of renin, angiotensin converting enzyme and angiotensin II levels in human placenta, chorion and amnion from women with pregnancy induced hypertension. Clinical Endocrinology (Oxf). 1996;**44**:429-433

[89] Pringle KG, Tadros MA, Callister RJ, Lumbers ER. The expression and localization of the human placental prorenin/renin-angiotensin system throughout pregnancy: Roles in trophoblast invasion and angiogenesis? Placenta. 2011;**32**:956-962. DOI: 10.1016/j.placenta.2011.09.020

[90] Lumbers ER, Wang Y, Delforce SJ, Corbisier de Meaultsart C, Logan PC, Mitchell MD, et al. Decidualisation of human endometrial stromal cells is associated with increased expression and secretion of prorenin. Reproduction of Biological Endocrinology. 2015;**13**:129. DOI: 10.1186/s12958-015-0127-8

[91] Pepin E, Dehboneh SS, Raguema N, Esfandarani MT, Lavoie JL. Role of the renin-angiotensin system in healthy and pathological pregnancies, renin-angiotensin system—Past, present and future. In: Anna Naidenova Tolekova. IntechOpen; 2017. DOI: 10.5772/66748

[92] Khlestova GV, Romanov AY, Nizyaeva NV, Karapetyan AO, Baev OR, Ivanets TY. Dynamics of renin, angiotensin II, and angiotensin (1-7) during pregnancy and predisposition to hypertension-associated complications. Bulletin of Experimental Biology and Medicine. 2018;**165**:438-439. DOI: 10.1007/s10517-018-4188-5

[93] Jeyabalan A, Conrad KP. Renal function during normal pregnancy and preeclampsia. Frontiers in Bioscience. 2007;**12**:2425-2437

[94] Robb CA, Davis JO, Johnson JA, Blaine EH, Schneider EG, Baumber JS. Mechanisms regulating the renal excretion of sodium during pregnancy. The Journal of Clinical Investigation. 1970;**49**:871-880. DOI: 10.1172/JCI106306

[95] Satué K, Domingo R. Longitudinal study of the renin angiotensin aldosterone system in purebred Spanish broodmares during pregnancy. Theriogenology. 2011;**75**(7):1185-1194. DOI: 10.1016/j.theriogenology.2010.11.029

[96] Laragh JH, Sealey JE. The plasma renin test reveals the contribution of body sodium-volume content (V) and renin-angiotensin (R) vasoconstriction

to long-term blood pressure. American Journal of Hypertension. 2011;**24**(11):1164-1180. DOI: 10.1038/ajh.2011.171

[97] Brosnihan AL. Systemic and uteroplacental renin–angiotensin system in normal and pre-eclamptic pregnancies. Therapeutic Advances in Cardiovascular Disease. 2008;**2**:349-362. DOI: 10.1177/1753944708094529

[98] Satué K, Domingo R, Redondo JI. Relationship between progesterone, oestrone sulphate and cortisol and the components of renin angiotensin aldosterone system in Spanish purebred broodmares during pregnancy. Theriogenology. 2011;**76**(8):1404-1415. DOI: 10.1016/j.theriogenology.2011.06.009

[99] Weir RJ, Brown JJ, Fraser R, Lever AF, Logan RW, McIlwaine GM, et al. Relationship between plasma renin, renin substrate, angiotensin II, aldosterone and electrolytes in normal pregnancy. The Journal of Clinical Endocrinology and Metabolism. 1975;**40**:108-115. DOI: 10.1210/jcem-40-1-108

[100] Symonds EM. The renin angiotensin system in pregnancy. Obstetrics & Gynecology Annual. 1981;**10**:45-67

[101] Van Dijk DJ, Boner G, Giler S, Erman A. Increased serum angiotensin converting enzyme activity and plasma angiotensin II levels during pregnancy and postpartum in the diabetic rat. JRAAS. 2001;**2**:193-198. DOI: 10.3317/jraas.2001.027

[102] Godard C, Gaillard R, Vallotton MB. The renin angiotensin aldosterone system in mother and fetus at term. Nephron. 1976;**17**:353-360. DOI: 10.1159/000180741

[103] McKeever KH, Hinchcliff KW, Schmall LM, Reed SM, Lamb DR, Muir W III. Plasma renin activity, aldosterone, and vasopressin during incremental treadmill exercise in horses. American Journal of Veterinary Research. 1992;**53**:1290-1293

[104] McCance DR, McKnight JA, Traub AI, Sheridan B, Roberts G, Atkinson AB. Plasma atrial natriuretic factor levels during normal pregnancy and pregnancy complicated by diabetes mellitus and hypertension. Journal of Human Hypertension. 1990;**4**:31-35

[105] Brochu M, Gauvin JP, St. Louis J. Increase of aldosterone secretion in adrenal cortex suspensions derived from pregnant rats. Proceedings of the Society for Experimental Biology and Medicine. 1996;**212**:147-152

[106] Brochu M, Lehoux JG, Picard S. Effects of gestation on enzymes controlling aldosterone synthesis in the rat adrenal. Endocrinology. 1997;**138**:2354-2358. DOI: 10.1210/endo.138.6.5198

[107] Brochu M, Roy-Clavel E, Picard S, St-Louis J. In vivo regulation of enzymes controlling aldosterone synthesis in pregnant rats. The Journal of Endocrinology. 1998;**24**:575-579

[108] Abou-Samra AB, Pugeat M, Dechaud H. Increased plasma concentration of N terminal beta lipotrophin and unbound cortisol during pregnancy. Clinical Endocrinology. 1984;**20**:221-228

[109] Jensen E, Wood CH, Keller-Wood M. The normal increase in adrenal secretion during pregnancy contributes to maternal volume expansion and fetal homeostasis. Journal of the Society for Gynecologic Investigation. 2002;**9**:362-371

[110] Wilson M, Morganti AA, Zervoudakis I, Letcher RI, Romns BM, Von Oeyon P, et al. Blood pressure, the renin aldosterone system and sex

steroids throughout normal pregnancy. The American Journal of Medicine. 1980;**68**:97-104

[111] Petit A, Guillon G, Tence M, Jard S, Gallo-Payet N, Bellabarba D, et al. Angiotensin II stimulates both inositol phosphate production and human placental lactogen release from human trophoblastic cells. The Journal of Clinical Endocrinology and Metabolism. 1989;**69**(2):280-286

[112] West CA, Sasser JM, Baylis C. The enigma of continual plasma volume expansion in pregnancy: Critical role of the renin-angiotensin-aldosterone system. American Journal of Physiology. Renal Physiology. 2016;**311**:1125-1134. DOI: 10.1152/ajprenal.00129.2016

[113] Hayslett JP. A new role for progesterone: An agonist for mineralocorticoid receptor activation and pregnancy-related hypertension. The American Journal of Kidney Diseases. 2001;**38**:893-895. DOI: 10.1053/ajkd.2001.27723

[114] Bay W, Ferris T. Factors controlling plasma renin and aldosterone during pregnancy. Hypertension. 1979;**1**:410-415

[115] Smeaton TC, Andersen GJ, Fulton JS. Study of aldosterone levels in plasma during pregnancy. The Journal of Clinical Endocrinology and Metabolism. 1977;**44**:1-7. DOI: 10.1210/jcem-44-1-1

Chapter 7

Connexin-Based Channels and RhoA/ROCK Pathway in Angiotensin II-Induced Kidney Damage

Gonzalo I. Gómez, Victoria Velarde and Juan C. Sáez

Abstract

The incidence of chronic kidney diseases is increasing worldwide, and there is no efficient therapy to reduce this phenomenon. The main therapies currently available focus on the control of blood pressure and the optimization of the blockade of the renin-angiotensin system (RAS). In addition, it is known that in several models of kidney damage, the amounts of connexins are altered. On the other hand, fasudil, a selective ROCK blocker, has shown renoprotective effects. The beneficial effects of blocking the RhoA/ROCK pathway in renal function may be related to its action of reducing macrophage infiltration, inflammation, and oxidative stress (OS), its expression of extracellular matrix genes and proteinuria, or to its effects on connexin abundance. Even though a correlation has been found between renal damage, caused by an increase in the RAS activity, connexins, and the RhoA/ROCK signaling pathway, it has not yet been possible to clearly determine its functional significance. Moreover, it has not been possible to identify the preponderance of this signaling pathway in the development of chronic kidney diseases. Here, we describe the advances in this subject.

Keywords: hypertensive nephropathy, oxidative stress, fibrosis, inflammation, Cx43, fasudil

1. Introduction

Chronic kidney disease (CKD) is a worldwide public health problem whose prevalence is persistently increasing. It is estimated that about 10% of adults in developed countries suffer some degree of kidney damage [1]. Patients with CKD usually develop a progressive kidney damage characterized by glomerular sclerosis and/or tubulointerstitial fibrosis, which eventually leads to end-stage renal disease, the last stage of this condition [2]. The detrimental effect of this process includes the progressive reduction of glomerular filtration rate (GFR) given by an increase in damaged nephrons, which eventually leads to organ failure [3]. CKD has different etiologies, including diabetic nephropathy, hypertensive nephrosclerosis, and glomerulonephritis. However, regardless of the initial cause, the morphological characteristics, such as tubular necrosis and glomerular sclerosis, are similar [4, 5]. This condition induces the partial destruction of nephrons and the progressive failure of renal function [4, 5].

2. Hypertensive nephropathy

Hypertension is the second cause of end-stage renal disease (ESRD) [6, 7]. Hypertensive nephropathy starts in the glomeruli due to an increase in intraglomerular pressure. These initial events activate and damage mesangial cells, epithelial cells, and podocytes within the glomerulus. In turn, these cells produce vasoactive and pro-inflammatory mediators, which increase cell damage and favor fibrosis, reducing renal blood flow and glomerular filtration rate [8]. The renal corpuscle, formed by Bowman's capsule and glomerulus, is the fundamental structure in the filtration process. The glomerulus is formed mainly by blood capillaries, podocytes, and the mesangium. The mesangium plays a key role in the structural and functional stability of the glomerulus, allowing it to successfully fulfill its filtering function [9]. The mesangial cells (MCs) constitute 30–40% of the cellular population of the glomerulus, and their function is to support the glomerulus and participate in the maintenance of the opening of its capillaries, regulation of the glomerular filtration rate, and synthesis and degradation of extracellular matrix proteins [9].

The renin-angiotensin system (RAS) is the prototype of a classic systemic endocrine network whose actions in the kidney and adrenal gland include regulation of blood pressure, intravascular volume, and electrolyte balance [10]. The RAS plays an integral role in the homeostatic control of arterial pressure, tissue perfusion, and extracellular volume. This pathway is initiated by the regulated secretion of renin from the kidney, the rate-limiting processing enzyme [11]. RAS begins with the biosynthesis of renin by the juxtaglomerular (JG) cells. Active renin secretion is regulated mainly by (1) the renal baroreceptor mechanism in the afferent arteriole that senses changes in renal perfusion pressure; (2) changes in delivery of NaCl (sensed as changes in Cl concentration) to the macula densa cells of the distal tubule, which lie close to the JG cells and form the JG apparatus; (3) sympathetic nerve stimulation via beta-1 adrenergic receptors; and (4) negative feedback by a direct action of angiotensin II (AngII) on the JG cells [11]. Renin secretion is stimulated by a fall in perfusion pressure or in NaCl delivery and by an increase in sympathetic activity [11, 12]. Angiotensinogen is secreted constitutively by the liver and reacts with renin, ending transformed into the inactive decapeptide angiotensin I (AngI) [11]. AngI is hydrolyzed by angiotensin-converting enzyme (ACE), which removes the C-terminal dipeptide to form AngII, a potent vasoconstrictor [11, 12]. AngII is the primary effector of a variety of RAS-induced physiological and pathophysiological actions [11]. AngII, via the AT1 receptor, stimulates the production of aldosterone by the zona glomerulosa in the adrenal gland [11]. Aldosterone is a major regulator of sodium and potassium ion (Na^+ and K^+, respectively) balance and thus plays a major role in regulating extracellular volume [11, 12]. It enhances the reabsorption of Na^+ and water in the distal tubules and collecting ducts (as well as in the colon and salivary and sweat glands) and thereby promotes K^+ (and hydrogen ion) excretion [11, 12].

The vasoconstriction and the increase in blood pressure mediated by AngII represent only part of the pleiotropic actions of this peptide. AngII stimulates aldosterone secretion, cell infiltration, proliferation and migration, thrombosis, superoxide ion production, and other factors involved in nephropathy [8]. When MCs are stimulated with AngII, the synthesis of extracellular matrix is increased and accumulates in the extracellular space [13]. Activated MCs produce more reactive oxygen species (ROS) [14, 15] and synthesize and release more pro-inflammatory cytokines such as interleukin-1β (IL-1β) and tumor necrosis factor-α (TNF-α) and chemokines, such as the macrophage chemoattractant protein (MCP-1) and transforming growth factor β (TGF-β) [13, 15–17]. In addition, high concentrations of AngII maintained for long periods of time in mice induce an inflammatory response characterized by the expression of pro-inflammatory cytokines such as IL-1β and

TNF-α [18], infiltration of macrophages (positive ED-1) [19], tubular overexpression of osteopontin (OPN) [19], and the expression of other pro-inflammatory cytokines, chemokines, and adhesion molecules [8, 20]. AngII also increases the expression of nicotinamide adenine dinucleotide phosphate oxidase (NOX), one of the main enzymes in the generation of ROS, contributing to the onset of oxidative stress (OS), independent of the action of pro-inflammatory cytokines [21, 22]. The sum of these alterations leads to an activation of the transcription factor nuclear factor-NF-κB, which increases the synthesis and release of extracellular matrix protein, such as type IV collagen, laminin, and fibronectin [13, 16], inducing the formation of mesangial nodules with lesions that extend to the interstitial areas, hindering the adequate function of the glomerulus [9, 13, 15, 16]. In summary, intrarenal RAS is an important factor in the pathophysiology of hypertension and hypertensive nephropathy [23].

2.1 Angiotensin II and RhoA/ROCK pathway in renal damage

Two receptors, AngII II receptor type 1 (AT1R) and type 2 (AT2R), both coupled to different G proteins, mediate the actions of AngII. The AT1 receptor activates small G proteins, including Ras, Rac1, RhoA, and the Rho kinase system (ROCK) [24], while the AT2 receptor inhibits RhoA [25]. The Rho family of small GTPases (Rho GTPase) is constituted by monomeric G proteins of 20–40 kDa considered as molecular switches, which cycle between two conformational states, an active state bound to GTP and an inactive state bound to GDP. In mammals, this family is composed of 20 members, of which the most studied are Rac1, Cdc42, and RhoA. The latter being the most studied member of this family [26]. ROCK, an effector downstream of RhoA, is a serine–threonine kinase of around 160 kDa, which in mammals is present in two isoforms, ROCK1 and ROCK2 [27, 28]. ROCK is composed of an amino terminal kinase domain, followed by a super-coiled helix region, which contains the Rho-GTP binding site and a carboxy-terminus, which contains an internal domain rich in cysteine residues [27]. ROCK1 and ROCK2 are highly homologous, sharing an identity of approximately 65% in their amino acid sequences and approximately 92% homology in their amino terminal kinase domain [27, 28].

The RhoA/ROCK pathway has received considerable attention because of its implication in a wide variety of pathophysiological states present in cardiovascular diseases, pulmonary hypertension, Alzheimer's disease, and glaucoma [27]. The RhoA/ROCK pathway plays an important role in renal pathophysiology, where RhoA/ROCK participates in the regulation of pro-inflammatory cytokines (e.g., TNF-α and IL-1β) [24, 29] and increases the amount of TGF-β and NFκB [27]. On the other hand, great interest has been generated in the use of fasudil, a selective ROCK inhibitor, as regulator in a wide variety of animal models of kidney damage, including unilateral ureteral obstruction, hypertensive glomerulosclerosis, acute renal failure induced by ischemia–reperfusion or by contrast-induced acute kidney injury, and renal failure induced by AngII [27, 30]. Fasudil, a ROCK inhibitor, prevents kidney damage by reducing the expression of extracellular matrix genes, OS, pro-inflammatory cytokines, and macrophage infiltration and inhibiting the cascade of events that leads to these effects. Thus, both RhoA and ROCK could be considered as therapeutic targets to prevent hypertension and kidney damage [24, 27, 29, 31].

2.2 Connexins in hypertensive nephropathy

Gap junctions (GJs) are conglomerates of intercellular channels that result from docking of two HCs or connexins, each one contributed by one of the cells in contact and formed by six connexins (Cxs). GJs allow direct ion exchange (explaining the electrical coupling), small metabolites (e.g., nicotinamide adenine dinucleotide: NAD^+, glucose, lactate, and glutamate), and second messengers (e.g., cyclic adenosine monophosphate [cAMP] and inositol trisphosphate [IP_3]) between adjacent cells (explaining metabolic coupling) [32, 33]. There are 21 isoforms of Cxs in humans and 20 in rodents. Each Cx is named according to its approximate molecular weight, and its transmembrane structure identifies four transmembrane domains connected by two extracellular loops and one intracellular loop, which implies that the C and N terminals are in the intracellular space [34]. HCs are fundamental pathways for the exchange of ions and small molecules between the intra- and extracellular compartments. These structures can be opened under physiological and pathophysiological conditions, allowing the release of paracrine and autocrine signaling molecules to the extracellular medium (e.g., adenosine triphosphate [ATP], glutamate, NAD^+, and prostaglandin E2 [PGE2]) [35]. Each type of channel formed by the Cxs has unique gate properties, in addition to a characteristic conductance and permeability features [36–38].

The kidney regulates blood pressure mainly through excretion of Na^+ and water, depending on the hormonal action of the RAS and other hormone systems with renal action [39]. However, in order to fulfill this function, the kidney requires the coordinated action of different cell types, including vascular and tubular cells [39]. This intrarenal coordination has not yet been well established, but, since connexins (Cxs) are present in the kidney, the existence of direct communication between the different cell types of the nephron has been proposed to occur through gap junctions (GJs) and/or hemichannels (HCs) [39]. In mammalian kidneys, nine types of Cxs have been detected (Cx26, Cx30.3, Cx31, Cx32, Cx37, Cx40, Cx43, Cx45, and Cx46), which are located in the vasculature or in different segments of the renal tubule, where most likely fulfill different functions [39].

In cortical astrocytes, it has been demonstrated that two pro-inflammatory cytokines, TNF-α and IL-1β, reduce intercellular communication mediated by GJs and increase the permeability of the membrane through HCs formed by Cx43 (Cx43 HCs) [40]. This opposed regulation of Cx43 GJs and HCs also occurs in cultures of proximal tubule cells treated with metabolic inhibitors or pro-inflammatory cytokines, where an increase in the activity of Cx HCs has been demonstrated in response to these stimuli [41, 42]. In pathological conditions such as hypertension, the amount of renal Cxs is altered. For example, in the two-kidney one-clip model (2K1C), an increase in the amount of Cx43 mRNA and protein in the glomerulus was observed [43]. Recently Oliveira et al. showed for the first time that bone marrow mononuclear cell (BMMC) transplantation in clipped kidney of the 2K1C rats significantly increased N-cadherin, E-cadherin, connexin40, and nephrin expression accompanied by improved renal morphology and function and decreased fibrosis [44]. This cell-based therapy, especially using the mononuclear cell fraction, has shown to improve regeneration of multiple tissues under pathological conditions [44]. A recent study provided evidence that both Cx40 and Cx37 participate in endothelial nitric oxide synthase (eNOS) regulation in vivo, where in mice subjected to the 2K1C procedure, the interaction of Cx40 and Cx37 with eNOS was enhanced, resulting in increased nitric oxide (NO) release [45]. Mice lacking Cx40 featured decreased levels of eNOS [45], and in different models of hypertension, Cx37 selectively participates from an altered expression of AT2R [46]. In addition, the amount of Cx43 is increased in inflammatory processes in damaged renal tubules and in interstitial cells in human kidneys [47]. Toubas et al. observed in three different models of CKD (i.e., the transgenic renin [$Ren^{+/+}$] model, the administration of antibodies against the glomerular basement membrane [α-GMB], and the unilateral obstructive uropathy) an increase in the amount of renal Cx43. Consequently, they postulated that this change in Cx43 was

DOI: http://dx.doi.org/10.5772/intechopen.87040

altered by the development of inflammation in the damaged kidney [36]. Therefore, Cx43 is considered a new mediator of renal disease involved in central processes of inflammation and fibrosis, while its inhibition even after the initiation of the disease attenuates renal damage and preserves renal function in animal models of vascular, tubular, and glomerular CKD [48]. Although renal tissue expresses several Cxs, only a few studies have described the involvement of GJs and HCs in kidney damage, and no signaling pathway has been clearly associated with these changes [36, 41, 42]. Therefore, the role of Cx-based channels in normal renal tissue or in the development and progression of kidney damage remains largely unknown.

2.3 Connexin43 and the RhoA/ROCK pathway participate in angiotensin II-induced kidney damage

The main therapies for CKD currently available focus on the control of blood pressure and the optimization of the blockade of the renin-angiotensin system (RAS) [49]. The renal afferent arterioles are primarily responsible for regulating preglomerular resistance, renal blood flow, and GFR. Elevated renal vascular resistance and preglomerular reactivity are observed in AngII-induced hypertension [50]. Although many systemic, neural, paracrine, and autoregulatory mechanisms contribute to afferent arteriolar dynamics, in AngII-dependent hypertension, a direct effect has been observed between the RhoA/ROCK pathway and the endogenous production of AngII [50]. In our studies we have observed that, although treatment with fasudil does not reduce systolic blood pressure (SBP), the establishment of irreversible renal damage is prevented (**Figure 1**, **Table 1**), reducing inflammation, OS, and fibrosis, and also kept the amount of Cx43 and phosphorylated myosin phosphatase target subunit-1 (MYPT-1) at normal levels [51] (**Figure 2**). We have also identified the timepoint when renal damage turns irreversible and, as such, independent of the cause [51]. We considered that kidney damage became irreversible after 4 weeks of treatment with AngII since SBP, inflammation, OS, fibrosis, the amount of Cx43, and phosphorylation status of MYPT-1 remained high even after 2 weeks of AngII withdrawal [51]. On the contrary, these parameters were reversed in animals infused with AngII for 3 or less weeks, which indicates that AngII can generate alterations that can be compensated by kidney tissue that was not affected by AngII and/or recovery thanks to the small regeneration capacity of kidney tissue [51].

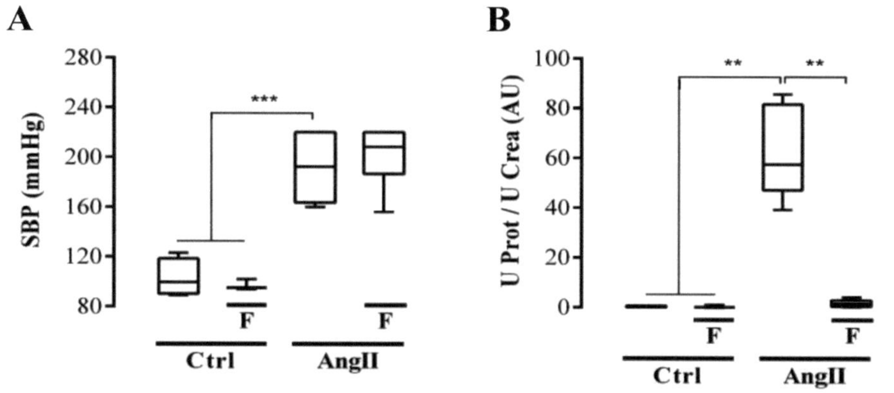

Figure 1.
*Fasudil does not modify the SBP in rats treated with AngII for 4 weeks but prevents the decrease in renal function. (A) Protein (UProt) and creatinine (UCrea) were measured in urine samples to assess renal function from ratio UProt/UCrea. (B) The bars represent the means ± SE of a n ≥ 4 rats per experimental group. The differences between the subgroups of each of the three groups were evaluated by an ANOVA followed by a Tuckey test. ***p < 0.001, **p < 0.01, and * p < 0.05 vs. AngII group.*

Groups	Weight (gr)	Proteinuria (mg/day)	Creatinine clearance (ml/min)	FE Na + (%)	FE K+ (%)
Ctrl	482 ± 31	2.7 ± 1.1***	1.4 ± 0.3***	0.2 ± 0.0***	12.0 ± 2.7***
Ctrl+fasudil	480 ± 36	3.6 ± 1.1***	2.1 ± 0.1***	0.1 ± 0.0***	12.5 ± 0.3***
AngII	364 ± 42	214.0 ± 19.0	0.7 ± 0.0	2.2 ± 0.4	162.0 ± 23.0
AngII+fasudil	368 ± 17	19 ± 7.2***	1.9 ± 0.2***	0.5 ± 0.1***	30 ± 7.2***

****p< 0.001 vs. AngII groups (n≥4/all groups).*

Table 1.
Values for weight, proteinuria, creatininuria, creatininemia, creatinine clearance, and fractional excretion (FE) for Na^+ and K^+ in the experimental groups.

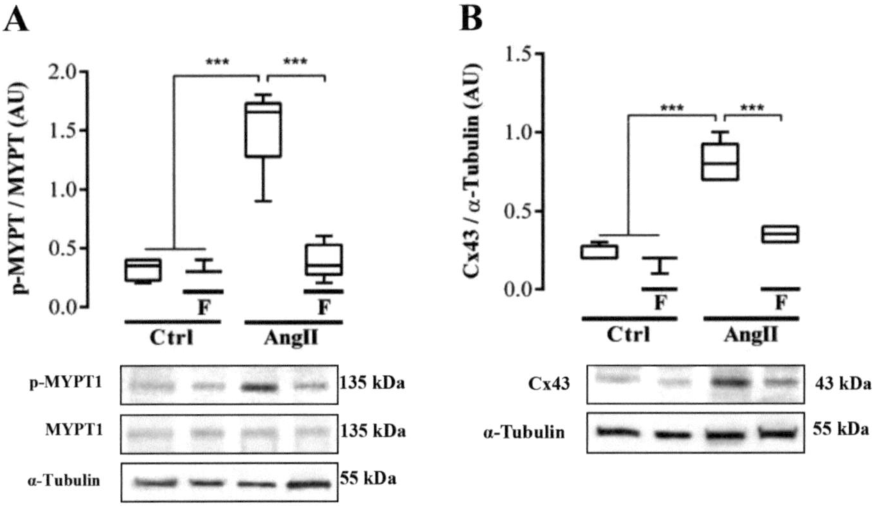

Figure 2.
*The increase in the amounts of phosphorylated MYPT and Cx43 is prevented with fasudil in rats treated with AngII for 6 weeks. Four groups of animals, two control groups (Ctrl and Ctrl+fasudil) and two experimental groups (AngII administered for 6 weeks and AngII+fasudil administered for the last 4 weeks), were studied. Fasudil (100 mg/kg/day) was given in the drinking water. Graphs show phosphorylation of MYPT-1 (A) and the relative amount of Cx43 (B). Under the graph representative pictures of phosphorylated MYPT (p-MYPT), unphosphorylated MYPT and Cx43 positive bands and its loading control (α-tubulin) are shown. The bars represent the means ± SE of n ≥ 4 rats per experimental group. The differences between the subgroups of each of the three groups were evaluated by an ANOVA followed by a Tuckey test. ***p < 0.001 vs. AngII group.*

The activity of the RhoA/ROCK pathway has been widely investigated in the pathogenesis of hypertension, where this pathway would fulfill an important role in the regulation of smooth muscle contraction. Other cellular processes such as proliferation, hypertrophy, adhesion, and migration of vascular cells are also mediated by the RhoA/ROCK pathway. These changes could lead to an increase in peripheral vascular resistance, which is one of the critical characteristics of several models of hypertension [52]. Therefore, inhibition of the RhoA/ROCK pathway represents a new approach in the prevention and treatment of hypertension [52]. The protective effect of fasudil in vivo is partly explained by its pleiotropic action in different systems. Therefore, considering that ROCK inhibitors were developed as antihypertensive drugs, it is striking that in our model of rats treated with AngII for 6 weeks, fasudil did not affect SBP, but did reduced the progression of kidney damage [51]. Similar to our observations, several studies have established that fasudil is renoprotective without affecting blood pressure, establishing a controversy regarding the use of fasudil and its antihypertensive action [53–56]. In view of these results, it would be interesting to develop a line of research that could explain why fasudil does not prevent the increase in SBP, even when it prevents kidney damage.

Hypertensive nephropathy begins in the glomerulus by increasing intraglomerular pressure. These early events activate and damage mesangial cells, epithelial cells, and podocytes in the glomerulus [42]. In turn, these cells produce vasoactive and pro-inflammatory agents, which increase cell damage and promote fibrosis, reducing renal blood flow and glomerular filtration [8]. In afferent arteriolar cells from rats treated with AngII, the activation of NF-κB is mediated by the RhoA/ROCK pathway, and the ROCK/NF-κB axis contributes to the upregulation of angiotensinogen, leading to an increase in the amount of intrarenal AngII [50]. We found that AngII increases the membrane permeability of MES cells, a mesangial glomerular cell line via AT1 receptors, as well as the activation of a RhoA/ROCK-dependent intracellular signaling pathway, followed by the upregulation of three nonselective channels, and the generation of OS and pro-inflammatory cytokines [42]. In MES-13 cells, AngII promotes a feedforward mechanism in which three nonselective channels (Cx43 HCs, Pannexin 1 channels, and $P2X_7$ receptors) maintain or even amplify inflammatory and oxidative responses, causing damage to kidney cells [42].

Xie et al. explored the mechanism of the reduction in the amount of Cx43 induced by RhoA/ROCK signaling in high glucose-treated glomerular mesangial cells (GMCs) [57]. Their results indicate that activated RhoA/ROCK signaling induced Cx43 degradation in GMCs cultured in high glucose via a pathway dependent on F-actin regulation that promoted the association between ZO-1 and Cx43 [57]. Interestingly, we found changes in RhoA/ROCK activity and also found that ROCK inhibitors prevented increases in the amount of Cx43 induced by AngII [51]. Since the expression and activation of RhoA/ROCK and Cx43 HCs, respectively, occur in the same direction, it is likely that they are regulated by the same transduction mechanism and intracellular signaling pathway activated by AngII. Therefore, it was postulated that changes in RhoA/ROCK pathway and Cx43 precede renal damage in this model of hypertensive nephropathy [51]. A comparable response has been found in fibroblasts, and a direct relationship has been demonstrated between the activation of the RhoA/ROCK pathway and the increase in the amount of Cx43. In these cells, the expansion mechanisms in response to stretching involve the release of ATP to the extracellular medium through the RhoA/ROCK pathway and the activation of Cx HCs [58]. In addition, treatment with Y-27632, another inhibitor of the RhoA/ROCK pathway, or with blockers of Cx HCs, such as octanol or carbenoxolone, inhibits the increase of ATP in the extracellular medium and the growth of fibroblast [58]. Nevertheless, this direct relationship is not observed in all cell types. For instance, in corneal epithelial cells, where a RhoA/ROCK-dependent pathway is involved in the formation of Cx43 GJs, inhibition of RhoA/ROCK-dependent pathway results in greater cell–cell communication mediated by Cx43 GJs [59].

Therefore, we propose that blocking AngII-induced damage progression in mesangial cell could be accomplished by inhibiting the RhoA/ROCK as previously demonstrated. Moreover, the effective reduction of initial AngII-induced alterations in cell membrane permeability leading to activation of several metabolic pathways that promote OS and generation of pro-inflammatory cytokines can be accomplished with selective and potent inhibitors of nonselective channels [42, 51].

The differences between the subgroups in each of the three groups were evaluated by an ANOVA followed by a Tuckey test. $^{**}p < 0.01$, $^{***}p < 0.001$ vs. AngII groups ($n \geq 4$/all groups).

3. Conclusions

In conclusion in the hypertensive nephropathy, inflammation, oxidative stress, fibrosis, changes in amount and cell membrane permeability of Cx43 HCs, and activity of the RhoA/ROCK pathway are important in the progression of damage induced by AngII. These alterations are prevented by fasudil, revealing a close relationship between activation of a RhoA/ROCK-dependent pathway and Cx43 in CKD.

Acknowledgements

The author would like to thank CONICYT, Fondecyt, Universidad Autónoma de Chile, and Pontificia Universidad Católica de Chile. This work was partially supported by a CONICYT Ph.D. fellowship No. 21120081 (Gonzalo I. Gómez) and a FONDECYT grant No. 1150291 (Juan C. Sáez) and ICM-Economía P09-022-F (Juan C. Sáez).

Conflict of interest

The author declares that the research was conducted in the absence of any commercial or financial relationships that could be construed as a potential conflict of interest.

Notes/thanks/other declarations

Gonzalo I. Gómez would like to thank my parents, sister, girlfriend, and teachers for their unconditional support, patience, and belief in me.

Appendices and nomenclature

CKD	chronic kidney diseases
RAS	renin-angiotensin system
ESRD	end-stage renal disease
GFR	glomerular filtration rate
MCs	mesangial cells
JG	juxtaglomerular cells
AngI	angiotensin I
ACE	angiotensin-converting enzyme
AngII	angiotensin II
IL-1β	interleukin-1β
TNF-α	tumor necrosis factor-α
MCP-1	macrophage chemoattractant protein
TGF-β	transforming growth factor-β
ED-1	infiltration of macrophages
OPN	osteopontin
NOX	nicotinamide adenine dinucleotide phosphate oxidase
ROS	reactive oxygen species
OS	oxidative stress
AT1	angiotensin II receptor type 1
AT2	angiotensin II receptor type 2

DOI: http://dx.doi.org/10.5772/intechopen.87040

ROCK	Rho kinase system
Rho GTPase	Rho family of small GTPases
NFκB	nuclear factor-κB
GJs	gap junctions
Cxs	connexins
HCs	hemichannels
BMMC	bone marrow mononuclear cell
2K1C	two-kidney and one-clip rat model
eNOS	endothelial nitric oxide synthase
SBP	systolic blood pressure
MYPT-1	myosin phosphatase target subunit-1
MES-13	mesangial glomerular cells line

Author details

Gonzalo I. Gómez[1]*, Victoria Velarde[2] and Juan C. Sáez[2,3]

1 Institute of Biomedical Science, Faculty of Health Science, Universidad Autónoma de Chile, Santiago, Chile

2 Department of Physiology, Faculty of Biological Science, Pontificia Universidad Católica de Chile, Santiago, Chile

3 Instituto de Neurociencias, Centro Interdisciplinario de Neurociencias de Valparaíso, Universidad de Valparaíso, Valparaíso, Chile

*Address all correspondence to: gonzalo.gomez@uautonoma.cl

References

[1] de Zeeuw D, Hillege HL, de Jong PE. The kidney, a cardiovascular risk marker, and a new target for therapy. Kidney International. Supplement [Internet]. 2005;**68**(98):S25-S29. [cited 12 December 2018]. Available from: http://linkinghub.elsevier.com/retrieve/pii/S008525381551260X

[2] Levey AS, Coresh J, Balk E, Kausz AT, Levin A, Steffes MW, et al. National Kidney Foundation practice guidelines for chronic kidney disease: Evaluation, classification, and stratification. Annals of Internal Medicine [Internet]. 2003; **139**(2):137-147. [cited 12 December 2018]. Available from: http://www.ncbi.nlm.nih.gov/pubmed/12859163

[3] López-Novoa JM, Martínez-Salgado C, Rodríguez-Peña AB, López-Hernández FJ. Common pathophysiological mechanisms of chronic kidney disease: Therapeutic perspectives. Pharmacology & Therapeutics [Internet]. 2010;**128**(1): 61-81. [cited 12 December 2018]. Available from: https://linkinghub.elsevier.com/retrieve/pii/S016372581000118X

[4] Coresh J, Byrd-Holt D, Astor BC, Briggs JP, Eggers PW, Lacher DA, et al. Chronic kidney disease awareness, prevalence, and trends among U.S. adults, 1999 to 2000. Journal of the American Society of Nephrology [Internet]. 2005;**16**(1):180-188. [cited 12 December 2018]. Available from: http://www.jasn.org/cgi/doi/10.1681/ASN.2004070539

[5] Nangaku M. Chronic hypoxia and tubulointerstitial injury: A final common pathway to end-stage renal failure. Journal of the American Society of Nephrology [Internet]. 2005;**17**(1):17-25. [cited 12 December 2018]. Available from: http://www.ncbi.nlm.nih.gov/pubmed/16291837

[6] Correa-Rotter R, Cusumano AM. Present, prevention, and management of chronic kidney disease in Latin America. Blood Purification [Internet]. 2008;**26**(1):90-94. [cited 12 December 2018]. Available from: http://www.ncbi.nlm.nih.gov/pubmed/18182804

[7] Xue JL, Ma JZ, Louis TA, Collins AJ. Forecast of the number of patients with end-stage renal disease in the United States to the year 2010. Journal of the American Society of Nephrology [Internet]. 2001;**12**(12):2753-2758. [cited 12 December 2018]. Available from: http://www.ncbi.nlm.nih.gov/pubmed/11729245

[8] Ozawa Y, Kobori H, Suzaki Y, Navar LG. Sustained renal interstitial macrophage infiltration following chronic angiotensin II infusions. American Journal of Physiology. Renal Physiology [Internet]. 2007;**292**(1):F330-F339. [cited 12 December 2018]. Available from: http://www.physiology.org/doi/10.1152/ajprenal.00059.2006

[9] Hernández-Salinas R, Vielma AZ, Arismendi MN, Boric MP, Sáez JC, Velarde V. Boldine prevents renal alterations in diabetic rats. Journal Diabetes Research [Internet]. 2013;**2013**:1-12. [cited 05 March 2019]. Available from: http://www.ncbi.nlm.nih.gov/pubmed/24416726

[10] Border WA, Noble NA. Interactions of transforming growth factor-beta and angiotensin II in renal fibrosis. Hypertens (Dallas, Tex 1979) [Internet]. 1998;**31**(1 Pt 2):181-188. [cited 12 December 2018]. Available from: http://www.ncbi.nlm.nih.gov/pubmed/9453300

[11] Atlas SA. The renin-angiotensin aldosterone system: Pathophysiological role and pharmacologic inhibition. Journal of Managed Care Pharmacy [Internet]. 2007;**13**(8 Supp B):9-20. [cited 22 April 2019]. Available from: http://www.ncbi.nlm.nih.gov/pubmed/17970613

[12] Funder JW. New biology of aldosterone, and experimental studies on the selective aldosterone blocker eplerenone. American Heart Journal [Internet]. 2002;**144**(5 Suppl):S8-S11. [cited 22 April 2019]. Available from: http://www.ncbi.nlm.nih.gov/pubmed/12422135

[13] Bolick DT, Hatley ME, Srinivasan S, Hedrick CC, Nadler JL. Lisofylline, a novel antiinflammatory compound, protects mesangial cells from hyperglycemia- and angiotensin II-mediated extracellular matrix deposition. Endocrinology [Internet]. 2003;**144**(12):5227-5231. [cited 05 March 2019]. Available from: https://academic.oup.com/endo/article-lookup/doi/10.1210/en.2003-0739

[14] Moon J-Y, Tanimoto M, Gohda T, Hagiwara S, Yamazaki T, Ohara I, et al. Attenuating effect of angiotensin-(1-7) on angiotensin II-mediated NAD(P)H oxidase activation in type 2 diabetic nephropathy of KK-A(y)/Ta mice. American Journal of Physiology. Renal Physiology [Internet]. 2011;**300**(6):F1271-F1282. [cited 05 March 2019]. Available from: http://www.physiology.org/doi/10.1152/ajprenal.00065.2010

[15] Yue L, Wang W, Wang Y, Du T, Shen W, Tang H, et al. Bletilla striata polysaccharide inhibits angiotensin II-induced ROS and inflammation via NOX4 and TLR2 pathways. International Journal of Biological Macromolecules [Internet]. 2016;**89**:376-388. [cited 05 March 2019]. Available from: https://linkinghub.elsevier.com/retrieve/pii/S014181301630407X

[16] Ding G-X, Zhang A-H, Huang S-M, Pan X-Q, Chen R-H. SP600125, an inhibitor of c-Jun NH2-terminal kinase, blocks expression of angiotensin II-induced monocyte chemoattractant protein-1 in human mesangial cells. World Journal of Pediatrics [Internet]. 2010;**6**(2):169-176. [cited 05 March 2019]. Available from: http://link.springer.com/10.1007/s12519-010-0033-2

[17] Sánchez-López E, Rodriguez-Vita J, Cartier C, Rupérez M, Esteban V, Carvajal G, et al. Inhibitory effect of interleukin-1beta on angiotensin II-induced connective tissue growth factor and type IV collagen production in cultured mesangial cells. American Journal of Physiology. Renal Physiology [Internet]. 2008;**294**(1):F149-F160. [cited 12 December 2018]. Available from: http://www.physiology.org/doi/10.1152/ajprenal.00129.2007

[18] Singh P, Bahrami L, Castillo A, Majid DSA. TNF-α type 2 receptor mediates renal inflammatory response to chronic angiotensin II administration with high salt intake in mice. American Journal of Physiology. Renal Physiology [Internet]. 2013;**304**(7):F991-F999. [cited 12 December 2018]. Available from: http://www.physiology.org/doi/10.1152/ajprenal.00525.2012

[19] Mezzano SA, Ruiz-Ortega M, Egido J. Angiotensin II and renal fibrosis. Hypertens (Dallas, Tex 1979) [Internet]. 2001;**38**(3 Pt 2):635-638. [cited 12 December 2018]. Available from: http://www.ncbi.nlm.nih.gov/pubmed/11566946

[20] Mezzano SA, Aros CA, Droguett A, Burgos ME, Ardiles LG, Flores CA, et al. Renal angiotensin II up-regulation and myofibroblast activation in human membranous nephropathy. Kidney International. Supplement [Internet]. 2003;**86**:S39-S45. [cited 12 December 2018]. Available from: http://www.ncbi.nlm.nih.gov/pubmed/12969126

[21] Rivera P, Ocaranza MP, Lavandero S, Jalil JE. Rho kinase activation and gene expression related to vascular remodeling in normotensive rats with high angiotensin I converting enzyme levels. Hypertens (Dallas, Tex 1979) [Internet]. 2007;**50**(4):792-798. [cited 12 December 2018]. Available from: https://www.ahajournals.org/doi/10.1161/HYPERTENSIONAHA.107.095117

[22] Zhao W, Chen SS, Chen Y, Ahokas RA, Sun Y. Kidney fibrosis in hypertensive rats: Role of oxidative stress. American Journal of Nephrology [Internet]. 2008;**28**(4):548-554. [cited 12 December 2018]. Available from: https://www.karger.com/Article/FullText/115289

[23] Johnson RJ, Alpers CE, Yoshimura A, Lombardi D, Pritzl P, Floege J, et al. Renal injury from angiotensin II-mediated hypertension. Hypertens (Dallas, Tex 1979) [Internet]. 1992;**19**(5):464-474. [cited 12 December 2018]. Available from: http://www.ncbi.nlm.nih.gov/pubmed/1568765

[24] Rupérez M, Sánchez-López E, Blanco-Colio LM, Esteban V, Rodríguez-Vita J, Plaza JJ, et al. The Rho-kinase pathway regulates angiotensin II-induced renal damage. Kidney International. Supplement [Internet]. 2005;**68**(99):S39-S45. [cited 12 March 2019]. Available from: https://linkinghub.elsevier.com/retrieve/pii/S0085253815512775

[25] Guilluy C, Rolli-Derkinderen M, Loufrani L, Bourgé A, Henrion D, Sabourin L, et al. Ste20-related kinase SLK phosphorylates Ser188 of RhoA to induce vasodilation in response to angiotensin II Type 2 receptor activation. Circulation Research [Internet]. 2008;**102**(10):1265-1274. [cited 12 December 2018]. Available from: https://www.ahajournals.org/doi/10.1161/CIRCRESAHA.107.164764

[26] Etienne-Manneville S, Hall A. Rho GTPases in cell biology. Nature [Internet]. 2002;**420**(6916):629-635. [cited 12 December 2018]. Available from: http://www.nature.com/articles/nature01148

[27] Kolavennu V, Zeng L, Peng H, Wang Y, Danesh FR. Targeting of RhoA/ROCK signaling ameliorates progression of diabetic nephropathy independent of glucose control. Diabetes [Internet]. 2008;**57**(3):714-723. [cited 12 December 2018]. Available from: http://diabetes.diabetesjournals.org/lookup/doi/10.2337/db07-1241

[28] Peng F, Wu D, Gao B, Ingram AJ, Zhang B, Chorneyko K, et al. RhoA/Rho-kinase contribute to the pathogenesis of diabetic renal disease. Diabetes [Internet]. 2008;**57**(6):1683-1692. [cited 12 December 2018]. Available from: http://diabetes.diabetesjournals.org/cgi/doi/10.2337/db07-1149

[29] Zhou L, Liu F, Huang XR, Liu F, Chen H, Chung ACK, et al. Amelioration of albuminuria in ROCK1 knockout mice with streptozotocin-induced diabetic kidney disease. American Journal of Nephrology [Internet]. 2011;**34**(5):468-475. [cited 12 December 2018]. Available from: https://www.karger.com/Article/FullText/332040

[30] Wang Y, Zhang H, Yang Z, Miao D, Zhang D. Rho kinase inhibitor, fasudil, attenuates contrast-induced acute kidney injury. Basic & Clinical Pharmacology & Toxicology [Internet]. 2018;**122**(2):278-287. [cited 08 May 2019]. Available from: http://doi.wiley.com/10.1111/bcpt.12895

[31] Kushiyama T, Oda T, Yamamoto K, Higashi K, Watanabe A, Takechi H, et al. Protective effects of Rho kinase inhibitor fasudil on rats with chronic kidney disease. American Journal of Physiology. Renal Physiology [Internet]. 2013;**304**(11):F1325-F1334. [cited 12 December 2018]. Available from: http://www.physiology.org/doi/10.1152/ajprenal.00447.2012

[32] Söhl G, Maxeiner S, Willecke K. Expression and functions of neuronal gap junctions. Nature Reviews. Neuroscience [Internet]. 2005;**6**(3):191-200. [cited 23 April 2019]. Available from: http://www.ncbi.nlm.nih.gov/pubmed/15738956

[33] Goodenough DA, Paul DL. Gap junctions. Cold Spring Harbor Perspectives in Biology [Internet]. 2009;**1**(1):a002576. [cited 23 April 2019]. Available from: http://cshperspectives.cshlp.org/lookup/doi/10.1101/cshperspect.a002576

[34] Abascal F, Zardoya R. Evolutionary analyses of gap junction protein families. Biochimica et Biophysica Acta [Internet]. 2013;**1828**(1):4-14. [cited 12 October 2018]. Available from: http://linkinghub.elsevier.com/retrieve/pii/S0005273612000582

[35] Gajardo-Gómez R, Labra VC, Orellana JA. Connexins and pannexins: New insights into microglial functions and dysfunctions. Frontiers in Molecular Neuroscience [Internet]. 2016;**9**:86. [cited 20 September 2018]. Available from: http://journal.frontiersin.org/Article/10.3389/fnmol.2016.00086/abstract

[36] Toubas J, Beck S, Pageaud A-L, Huby A-C, Mael-Ainin M, Dussaule J-C, et al. Alteration of connexin expression is an early signal for chronic kidney disease. American Journal of Physiology. Renal Physiology [Internet]. 2011;**301**(1):F24-F32. [cited 12 December 2018]. Available from: http://www.physiology.org/doi/10.1152/ajprenal.00255.2010

[37] Trosko JE, Chang C-C, Wilson MR, Upham B, Hayashi T, Wade M. Gap junctions and the regulation of cellular functions of stem cells during development and differentiation. Methods [Internet]. 2000;**20**(2):245-264. [cited 23 April 2019]. Available from: http://www.ncbi.nlm.nih.gov/pubmed/10671317

[38] Saez JC, Berthoud VM, Branes MC, Martinez AD, Beyer EC. Plasma membrane channels formed by connexins: Their regulation and functions. Physiological Reviews [Internet]. 2003;**83**(4):1359-1400. [cited 23 April 2019]. Available from: http://www.physiology.org/doi/10.1152/physrev.00007.2003

[39] Kurtz A. Renal connexins and blood pressure. Biochimica et Biophysica Acta [Internet]. 2012;**1818**(8):1903-1908. [cited 12 December 2018]. Available from: https://linkinghub.elsevier.com/retrieve/pii/S0005273611001714

[40] Retamal MA, Froger N, Palacios-Prado N, Ezan P, Sáez PJ, Sáez JC, et al. Cx43 hemichannels and gap junction channels in astrocytes are regulated oppositely by proinflammatory cytokines released from activated microglia. The Journal of Neuroscience [Internet]. 2007;**27**(50):13781-13792. [cited 12 December 2018]. Available from: http://www.jneurosci.org/cgi/doi/10.1523/JNEUROSCI.2042-07.2007

[41] Vergara L, Bao X, Cooper M, Bello-Reuss E, Reuss L. Gap-junctional hemichannels are activated by ATP depletion in human renal proximal tubule cells. The Journal of Membrane Biology [Internet]. 2003;**196**(3):173-184. [cited 12 December 2018]. Available from: http://link.springer.com/10.1007/s00232-003-0636-9

[42] Gómez G, Fernández P, Velarde V, Sáez J. Angiotensin II-induced mesangial cell damage is preceded by cell membrane permeabilization due to upregulation of non-selective channels. International Journal of Molecular Sciences [Internet]. 2018;**19**(4):957. [cited 07 October 2018]. Available from: http://www.ncbi.nlm.nih.gov/pubmed/29570626

[43] Haefliger J-A, Krattinger N, Martin D, Pedrazzini T, Capponi A, Döring B, et al. Connexin43-dependent mechanism modulates renin secretion and hypertension. The Journal of Clinical Investigation [Internet]. 2006;**116**(2):405-413. [cited 12 December 2018]. Available from: http://www.jci.org/cgi/doi/10.1172/JCI23327

[44] Oliveira M, Lira R, Freire T, Luna C, Martins M, Almeida A, et al. Bone marrow mononuclear cell transplantation rescues the glomerular filtration barrier and epithelial cellular junctions in a renovascular hypertension model. Experimental Physiology [Internet]. 2019;**104**(5):740-754. [cited 23 April 2019]. Available from: http://doi.wiley.com/10.1113/EP087330

[45] Abed AB, Kavvadas P, Chadjichristos CE. Functional roles of connexins and pannexins in the kidney. Cellular and Molecular Life Sciences [Internet]. 2015;**72**(15):2869-2877. [cited 23 April 2019]. Available from: http://link.springer.com/10.1007/s00018-015-1964-5

[46] Le Gal L, Pellegrin M, Santoro T, Mazzolai L, Kurtz A, Meda P, et al. Connexin37-dependent mechanisms selectively contribute to modulate angiotensin II-mediated hypertension. Journal of the American Heart Association [Internet]. 2019;**8**(8): e010823. [cited 08 May 2019]. Available from: https://www.ahajournals.org/doi/10.1161/JAHA.118.010823

[47] Hillis GS, Duthie LA, Mlynski R, McKay NG, Mistry S, MacLeod AM, et al. The expression of connexin 43 in human kidney and cultured renal cells. Nephron [Internet]. 1997;**75**(4):458-463. [cited 12 December 2018]. Available from: https://www.karger.com/Article/FullText/189585

[48] Prakoura N, Kavvadas P, Chadjichristos CE. Connexin 43: A new therapeutic target against chronic kidney disease. Cellular Physiology and Biochemistry [Internet]. 2018;**49**(3):985-985. [cited 08 May 2019]. Available from: http://www.ncbi.nlm.nih.gov/pubmed/30196298

[49] Vilayur E, Harris DCH. Emerging therapies for chronic kidney disease: What is their role? Nature Reviews. Nephrology [Internet]. 2009;**5**(7):375-383. [cited 19 December 2018]. Available from: http://www.nature.com/articles/nrneph.2009.76

[50] Miyata K, Satou R, Shao W, Prieto MC, Urushihara M, Kobori H, et al. ROCK/NF-κB axis-dependent augmentation of angiotensinogen by angiotensin II in primary-cultured preglomerular vascular smooth muscle cells. American Journal of Physiology. Renal Physiology [Internet]. 2014;**306**(6):F608-F618. [cited 19 December 2018]. Available from: http://www.physiology.org/doi/10.1152/ajprenal.00464.2013

[51] Gómez GI. In the model of rena damage induced by infusion with angiotensin II, the RhoA/ROCK pathway is activated and regulates the abundance and activity of the HCs and GJs formed by Cx43 [thesis]. Pontificia Universidad Católica de Chile. 2017

[52] Lai A, Frishman WH. Rho-kinase inhibition in the therapy of cardiovascular disease. Cardiology in Review [Internet]. 2005;**13**(6):285-292. [cited 19 December 2018]. Available from: http://www.ncbi.nlm.nih.gov/pubmed/16230885

[53] Kanda T, Wakino S, Hayashi K, Homma K, Ozawa Y, Saruta T. Effect of fasudil on Rho-kinase and nephropathy in subtotally nephrectomized spontaneously hypertensive rats. Kidney International [Internet]. 2003;**64**(6):2009-2019. [cited 19 December 2018]. Available from: http://linkinghub.elsevier.com/retrieve/pii/S0085253815495699

[54] Nishikimi T, Akimoto K, Wang X, Mori Y, Tadokoro K, Ishikawa Y, et al. Fasudil, a Rho-kinase inhibitor, attenuates glomerulosclerosis in Dahl salt-sensitive rats. Journal of Hypertension [Internet]. 2004;**22**(9):1787-1796. [cited 19 December 2018]. Available from: http://www.ncbi.nlm.nih.gov/pubmed/15311108

[55] Ishikawa Y, Nishikimi T, Akimoto K, Ishimura K, Ono H, Matsuoka H. Long-term administration of rho-kinase inhibitor ameliorates renal damage in malignant hypertensive rats. Hypertens (Dallas, Tex 1979) [Internet]. 2006;**47**(6):1075-1083. [cited 19 December 2018]. Available from: https://www.ahajournals.org/doi/10.1161/01.HYP.0000221605.94532.71

[56] Suzuki H, Yamamoto T, Fujigaki Y, Eguchi S, Hishida A. Comparison of ROCK and EGFR activation pathways in the progression of glomerular injuries in AngII-infused rats. Renal Failure [Internet]. 2011;**33**(10):1005-1012. [cited 08 April 2019]. Available from: http://www.tandfonline.com/doi/full/10.3109/0886022X.2011.618923

[57] Xie X, Chen C, Huang K, Wang S, Hao J, Huang J, et al. RhoA/rho kinase signaling reduces connexin43 expression in high glucose-treated glomerular mesangial cells with zonula occludens-1 involvement. Experimental Cell Research [Internet]. 2014;**327**(2):276-286. [cited 18 April 2019]. Available from: https://linkinghub.elsevier.com/retrieve/pii/S0014482714002997

[58] Langevin HM, Fujita T, Bouffard NA, Takano T, Koptiuch C, Badger GJ, et al. Fibroblast cytoskeletal remodeling induced by tissue stretch involves ATP signaling. Journal of Cellular Physiology [Internet]. 2013;**228**(9):1922-1926. [cited 19 December 2018]. Available from: http://doi.wiley.com/10.1002/jcp.24356

[59] Anderson SC, Stone C, Tkach L, Sundar Raj N. Rho and Rho-kinase (ROCK) signaling in adherens and gap junction assembly in corneal epithelium. Investigative Ophthalmology & Visual Science [Internet]. 2002;**43**(4):978-986. [cited 23 April 2019]. Available from: http://www.ncbi.nlm.nih.gov/pubmed/11923237

Chapter 8

The Potential Role of Renin Angiotensin System (RAS) and Dipeptidyl Peptidase-4 (DPP-4) in COVID-19: Navigating the Uncharted

Hayder M. Al-Kuraishy, Marwa S. Al-Niemi, Nawar R. Hussain, Ali I. Al-Gareeb, Nasser A. Al-Harchan and Azhar H. Al-Kurashi

Abstract

Novel coronavirus (COVID-19) led to infected pneumonia and acute respiratory distress syndrome (ARDS) and acute kidney injury (AKI). The entry-point receptor for COVID-19 is angiotensin-converting enzyme 2 (ACE2) at lung, and dipeptidyl peptidase-4 (DPP-4) is a receptor for Middle East respiratory syndrome coronavirus (MERS-CoV). There is 80% similarity between MERS-CoV and COVID-19. This study was planned to review the potential link between the incidence and severity of COVID-19 regarding the modulation of DPP-4 and ACE2 by DPP-4 and renin angiotensin system (RAS). In COVID-19, SARS-CoV2 binds ACE2 which is highly expressed by the epithelial cells of the blood vessel, intestine, and lung. However, pulmonary ACE2 seems to be a protective defense pathway during ARDS. DPP-4 is not concerned with the entry of COVID-19 but mediates the inflammatory reactions and cytokine storm that induced ARDS and AKI by COVID-19. The interaction between DPP4i and RAS inhibitors seem to augment the expression of AT2 receptor and ACE2 which are under extensive researches to find the pathophysiological pathway of COVID-19 infection. This beneficial interaction between DPP4i and RAS shed light for possible attenuation of COVID-19-induced ARDS and AKI mainly in critically ill patients with systemic hypertension.

Keywords: COVID-19, angiotensin-converting enzyme 2, dipeptidyl peptidase-4, acute respiratory distress syndrome, acute kidney injury

1. Introduction

Coronavirus (CoV) is a positive-sense single-strand RNA genome, which is the largest known RNA virus, constitutes *Riboviria* of *Coronaviridae* family and

Orthocoronavirinae subfamily with a genome size of 27–34 kilobases. The name coronavirus is derived from the Latin word corona, meaning crown or halo, which is the feature appearance under electron microscope. Regarding COVID-19, it is similar to the other RNA viruses with additional endoribonuclease and exoribonuclease terminal caps (**Figure 1**) [1].

Formerly different researches were done to identify the cause of obscure respiratory tract infections. The identified novel agents were severe acute respiratory syndrome coronavirus (SARS-CoV) and human coronavirus NL63 (HCoV-NL63) [2] CoV causes human acute respiratory infection in about 5–30% of the total respiratory infection. In the past, between 2002 and 2003, an epidemic had been shown caused by SARS-CoV, which causes severe acute respiratory syndrome in 8000 subjects with 750 deaths. SARS-CoV is a second coronavirus group, while HCoV-NL63 is called group 1 coronavirus [3, 4].

On December 2019, a number of cases with a novel coronavirus (COVID-19) led to infected pneumonia in Wuhan in central China. Four cases with pneumonia of unknown etiology were reported on 29 December 2019, which heralds the similar outbreak of 2003 SARS-CoV [5].

At this time, no vaccines or effective antiviral agents are approved for the treatment or prevention of human COVID-19. Therefore, the emergence of COVID-19 is regarded as a priority by the World Health Organization (WHO), Center for Disease Control and Prevention (CDC), and health agencies for the developing a therapeutic effective drug against COVID-19. In general COVID-19 as other positive-sense RNA virus is characterized by high genetic plasticity and mutations due to short replication time, with a higher rate of recombination. These criteria limit and challenge for discovering novel antiviral agents [6, 7]. However, nucleotide and nucleoside analogue inhibitors (NIs) may be an effective agent by the reduction of genetic mapping of COVID-19 [8]. NIs such as ribavirin, remdesivir, and beta-D-N4-hydroxycytidine could be effective agents despite emergence of rapid resistance [9].

Previously, it has been reported by Li et al. [11] and Kuba et al. [10] that the entry-point receptor for SARS-CoV is angiotensin-converting enzyme 2 (ACE2) at the lung [10, 11]. Besides, Eckerle et al. reported that dipeptidyl peptidase-4 (DPP-4) which cluster of differentiation 26 (CD26) is widely expressed on the surface of human cells which acts as a receptor for Middle East respiratory syndrome coronavirus (MERS-CoV) [12]. Since there is a 80% similarity between MERS-CoV and SARS-CoV with SARS-CoV-2, this study was planned to review

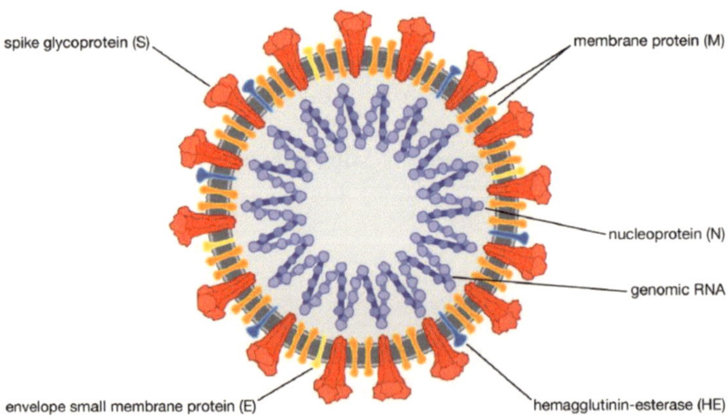

Figure 1.
Structure of SARS-CoV2.

the potential link between the incidence and severity of COVID-19 regarding the modulation of DPP-4 and ACE2 by DPP-4 and renin angiotensin system (RAS) inhibitors, respectively.

2. Literature search strategy

In general, an endeavor of this study article was to present a mini review concerning DPP-4 and RAS and their inhibitors in relation to the incidence and severity of COVID-19. Evidences and substantiations from experimental, preclinical, and clinical studies are assessed, given the nature and character of the subject area; it remains clear that this literature search cannot be considered as a systemic review. A multiplicity and array of search policy took on and are assumed, which is integrated by electronic database searches of Scopus, Web of Science, Medline, Cochrane Central Register of Controlled Trials (CCTR), and PubMed using MeSH terms, keywords, and title words during the search. The terms used for these searches were as follows: COVID-19 or coronavirus or SARS-CoV-2 and DPP-4 OR renin angiotensin system; COVID-19 or SARS-CoV-2 and DPP-4 inhibitors or sitagliptin, vildagliptin, and insulin resistance; incidence of COVID-19 or coronavirus or SARS-CoV-2; and hypertension or orphan drugs or glucagone-like peptide 1 (GLP-1) or dipeptidyl peptidase-IV inhibitors (DPPIV) or thiazolidinediones (TZDs).

Reference lists of previous and recent notorious articles were reviewed. In addition, only English articles were measured, and case reports were also of concerned in this review. The key features of predictable and suitable search studies were considered, and the conclusions were summarized and summed up in a mini review.

3. Renin angiotensin system (RAS) and COVID-19

RAS is a signaling pathway for the regulation of blood pressure, blood volume, natriuresis, and other vascular functions [13]. RAS consists of different effector peptides that control the dynamic vascular functions. Angiotensinogen from the liver is converted to angiotensin I (AngI), which is converted to angiotensin II (AngII) by angiotensin-converting enzyme 1(ACE1). AngII activates two types of receptors which are AT1 (vasoconstrictor) 90% and AT2 (vasodilator) 10%. The overall effect of AngII is vasoconstriction with sympathetic activation and aldosterone release. Excess of AngII is metabolized by ACE2 into vasodilator Ang (1–7) which act on specific receptor called MAS receptor (**Figure 2**). AngII can also be converted to angiotensin A by mononuclear leukocyte-derived aspartate decarboxylase (MLDAD), leading to the formation of alamandine, which has been shown to bind to the mas-related G protein-coupled receptor D [14].

In COVID-19, SARS-CoV-2 binds ACE2 which is highly expressed by the epithelial cells of the blood vessel, intestine, and lung. The expression of ACE2 is augmented by angiotensin-converting enzyme inhibitors (ACEIs) and angiotensin receptor blockers (ARBs) (**Figure 3**) [15] as well as ibuprofen and thiazolidinediones, and cigarette smoking also increases the expression of ACE2 at epithelial cells of the lung. Therefore, hypertensive patients on ACEIs or ARBs are at higher risk for COVID-19 [16, 17].

Recently, Zhou et al. [18] found that the COVID-19 genome is 96% identical to the bat CoV, so spike glycoprotein (S protein) and receptor binding domain (RBD) of both SARS-CoV-2 and bat CoV bind ACE2, which might explain the cross-species transmission, which means from bat to human and from human to human [19]. In addition, the affinity of COVID-19-RBD to ACE2 is approximately 10–20 times more

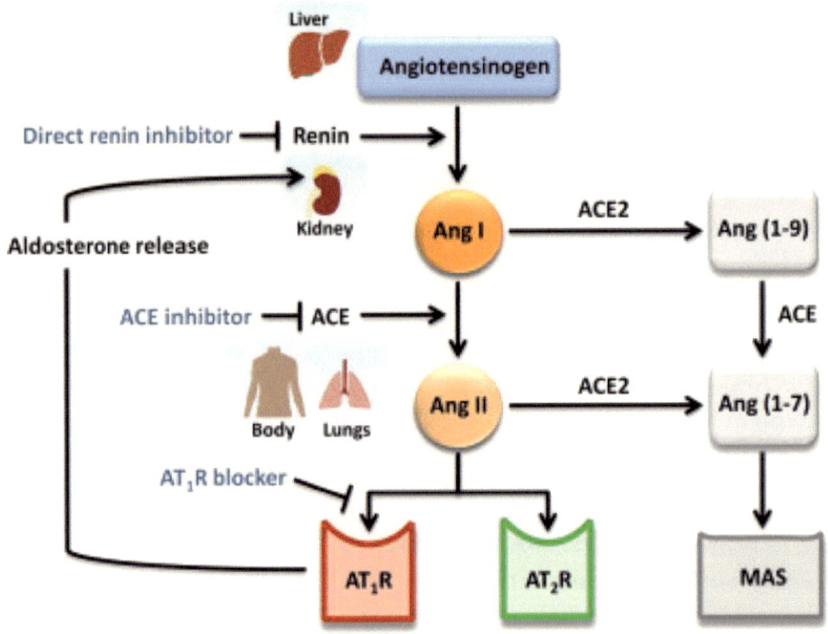

Figure 2.
Renin angiotensin system (RAS).

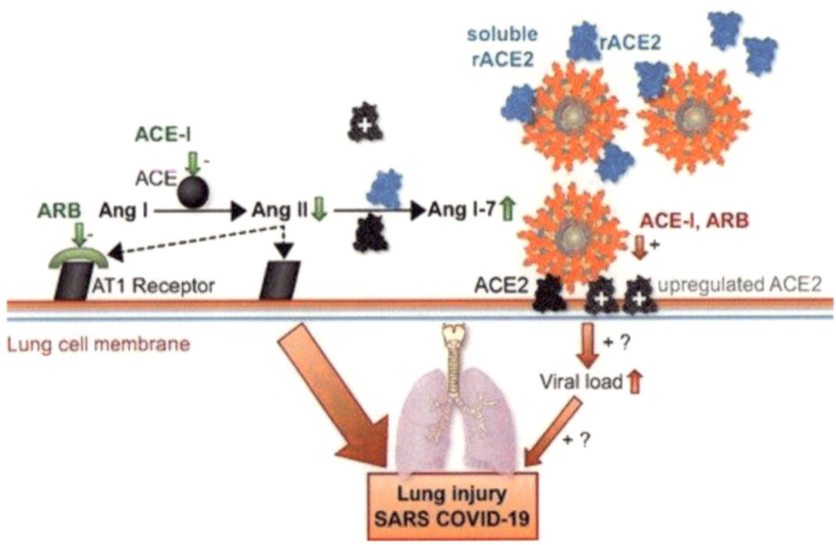

Figure 3.
Role of angiotensin-converting enzyme 2 (ACE2).

than that of SARS-CoV [20]. It has been shown that smoking upregulates ACE2 via activation of pulmonary platelet factor and induction of oxidative stress and inflammatory reactions. Increased ACE activity also contributes to impaired vascular relaxation observed in smokers. Irbesartan, an AT_1R antagonist (also known as angiotensin receptor blocker), was found to reduce arterial stiffness in hypertensive patients. When stratified by smoking status, smokers were found to have stiffer arteries before treatment, and irbesartan was able to reduce arterial stiffness to a greater extent in smokers than in nonsmokers, indicating that overactivity of the RAS contributes to increased arterial stiffness in smokers. On the other hand, overexpression and upregulation of ACE2 is regarded as a protective measure via the reduction of pro-inflammatory cytokines mainly tumor necrosis factor (TNF-α) which is augmented in smoker subjects [21]. Therefore, active cigarette smoking subjects are at high risk for COVID-19 regardless of gender difference (**Figure 4**).

The activity of RAS is high in the lung, which is the main source of circulating AngII due to higher expression of ACE. Lung ACE2 controls the balance of RAS activation through regulating AngII/Ang 1–7 ratio. Local pulmonary AngII provokes vascular permeability, causing pulmonary edema [22]. However, in acute respiratory distress syndrome (ARDS), the activation of RAS is necessary to maintain oxygenation since ACE2 knockout mice illustrated more sever pulmonary damage than the controls. Thereby, pulmonary ACE2 seems to be a protective defense pathway during ARDS [23]. In addition, ACE2 has an important anti-inflammatory action, and so ACE2 therapy is effective in the treatment of hypertension and diabetic nephropathy through attenuation of AngII-induced inflammation and oxidative stress [24]. Even so recombinant ACE2 is effective in the management of animal model ARDS, the inhibition of ACE2 may lead to fatal outcomes due to the reduction of vasodilator Ang1–7. However, chronic intravenous administration of Ang1–7 or MAS agonists leads to vasodilatation independent of circulating AngII levels [25].

Depending on these observations, different studies illustrate that RAS inhibitors might be of value in the reduction of ARDS, respiratory failure, and acute pneumonia that are induced by SARS-CoV-2 [26]. Though Wang [27] confirms that RAS inhibitors increase the risk of COVID-19 due to the upregulation of pulmonary ACE2, this study recommends stopping RAS inhibitors during COVID-19 outbreak. Nonetheless, all recruited patients with COVID-19 developed ARDS without any evidence of AKI [27]. Thus, RAS system mainly ACE2/Ang1–7 grows to be the focus and meeting point of different researches to implicate this pathway in the pathogenesis of COVID-19.

Guo et al. found that the expression of ACE2 is higher renal tubules than in lung tissues; nevertheless COVID-19 leads to ARDS in much at higher than that of acute

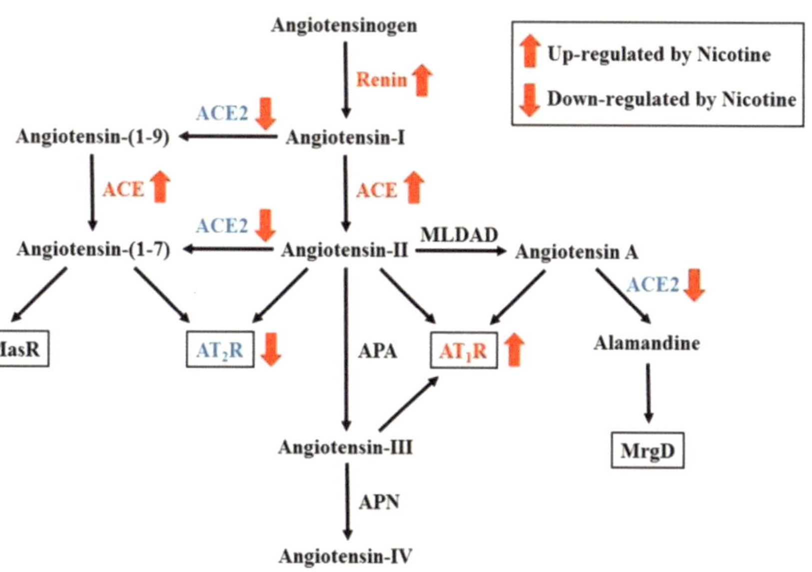

Figure 4.
Nicotine smoking and ACE2.

kidney injury (AKI), suggesting other mechanism other than ACE2 binding in the pathogenesis of COVID-19 [28]. AT2 receptor is activated by ACE2 and Ang1–7 that oppose the activity of AT1 receptor [29]. Similarly, AT2 receptors are highly expressed in lung epithelial cells compared with kidney tissues. Pulmonary AT2 receptors mediate lung injury through the augmentation of pulmonary inflammation and vascular permeability as well as the development of pulmonary fibrosis [30]. Consequently, pulmonary AT2 receptor antagonists are regarded as a novel pathway in COVID-19-induced pneumonia and ARDS. This finding does not rule out the responsibility of ACE2, since the activation of ACE2 by SARS-CoV-2 causes considerable activation of pulmonary AT2 receptors (**Figure 5**).

Amid myriad literature survey, polymorphism of ACE2 has been associated with different cardio-metabolic disorders; thus, the implication of ACE2 and AT2 receptors in COVID-19-induced pneumonia should be considerably regarded with ACE2 polymorphisms [31]. Furthermore, the expression of ACE2 might not be necessary for COVID-19 infection and viral entry, as the absence of SARS-CoV-2 in some ACE2 expressing cell types as well; this infection was observed in some cell line lacking ACE2, suggesting a vague pathway, and cofactors might be necessary for human infection [32]. Amusingly, Gurwitz [33] animal model study shows that pulmonary COVID-19 infection leads to significant lung injury through the downregulation of ACE2, which is attenuated by the administration of ARB. This study suggests the protective role of RAS inhibitors in COVID-19-induced ARDS [33]. Indeed, most of reported data from various social media during this dangerous outbreak was not legitimate due to overwhelming lay press and sparked concerns. Therefore, the high incidence of COVID-19 in patients receiving ACEIs or ARBs might be not because of these drugs but because those patients were often older, hypertensive, or diabetic which in fact increase the risk of COVID-19 infection. In addition, a recent clinical trial on the effectiveness of recombinant ACE2 (rACE) in the management of COVID-19 infection has been started [34], and we are waiting for these results. Thus, according to the guideline for the management of hypertension, RAS inhibitors should be used irrespective of COVID-19 infection, as sudden

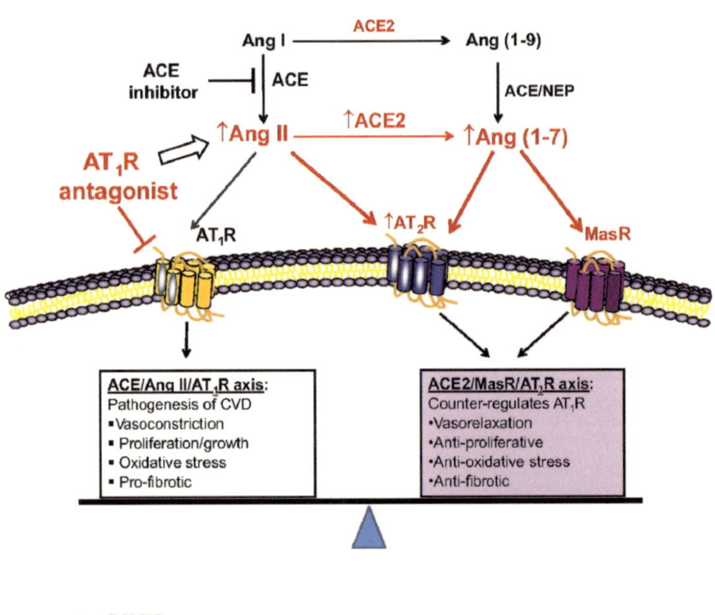

Figure 5.
Role of AT2 receptors in COVID-19.

withdrawal of these therapeutic regimens may increase the risk of deleterious outcomes in critically ill patients.

Regarding gender differences in the expression of ACE2, ACE2 gene is located on the X chromosome, which gives the possibility of gender differences in the susceptibility for COVID-19 infection. Females have lower levels of ACE2 compared with males, which gives a clue of male vulnerability to COVID-19 infection as compared with females [35]. In addition, Shenoy et al. report that estrogen attenuates AngII-induced pulmonary fibroblast proliferation due to the upregulation of

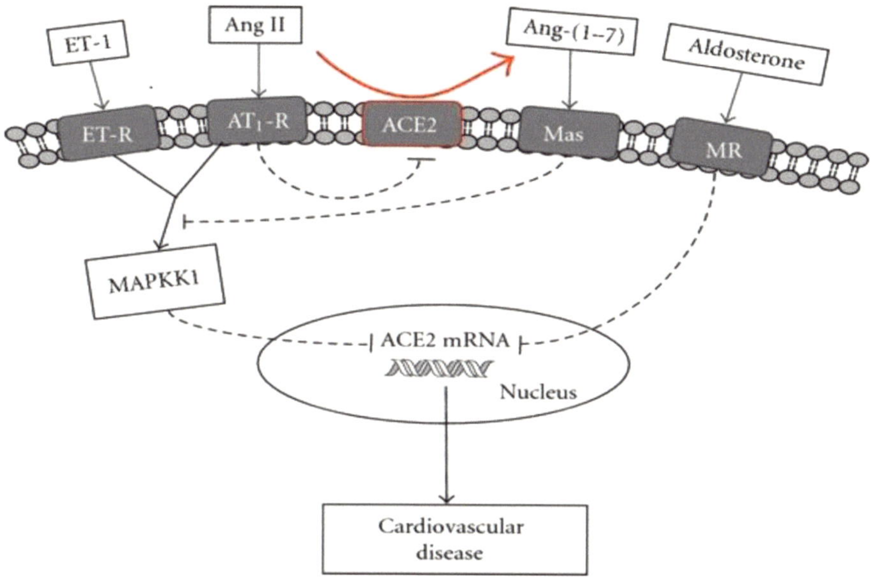

Figure 6.
Potential effects of endothelin-1 and aldosterone on the expression of ACE2.

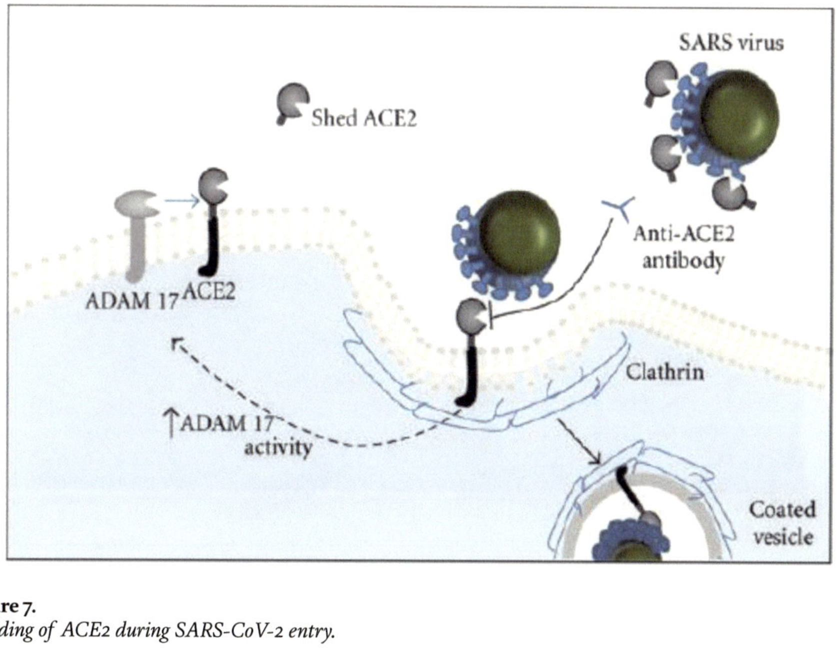

Figure 7.
Shedding of ACE2 during SARS-CoV-2 entry.

ACE2 [36]. The expression of ACE2 is regulated by different endogenous hormones and peptides; both endothelin-1 and aldosterone downregulate ACE2 expression in a rat model. So, RAS inhibitors may improve ACE2 expression via suppression of endogenous endothelin-1 and aldosterone (**Figure 6**) [37].

Moreover, ACE2 contains ectodomain and endodomain at cytoplasmic membrane; the shedding part (ectodomain) is essential for the replication of SARS-CoV-2 (**Figure** 7) [38].

At the heart of the dilemma, extensive researches are recommended to explore the specific role of ACE2 and RAS inhibitors during precarious COVID-19 worldwide outbreak.

4. Dipeptidyl peptidase-4 (DPP-4) and COVID-19

DPP4, also called adenosine deaminase complexing protein 2 (ADCP2) or adenosine deaminase binding protein (ADBP), is widely distributed on the surface of human cells. DPP4 is involved in the regulation of blood glucose, oxidative stress, inflammation, immune system, cell adhesion, and apoptosis [39]. The substrates of DPP4 are glucagone-like peptide 1 (GLP-1), stromal cell-derived factor 1 (SDF-1), brain natriuretic peptide (BNP), substance P, and neuropeptide Y (NPY). Furthermore, DPP4 interacts with different ligands such as caveolin-1, chemokine receptor type 4 (CXCR4), adenosine deaminase (ADA), and fibronectin [40].

DPP4 inhibitors (DPP4i) such as sitagliptin, saxagliptin, and vildagliptin are approved in the management of type 2 diabetes mellitus (T2DM). Additionally, DPP4i have a potential therapeutic effect regarding hypertension, endothelial dysfunction, cardioprotection, and connective tissue diseases (**Figure 8**) [41].

Regarding the role of DPP4 in viral entry and replication, Widagdo et al. [42] found that DPP4 is an important receptor for MERS-CoV transmission. DPP4 receptors are found in the human upper respiratory tract epithelium. Lacking of these receptors may restrict the transmission of MERS-CoV.

S protein of MERS-CoV binds specifically to DPP4 receptors. These interactions provoke proteolytic activation for viral entry and fusion of viral membrane with the cell membrane [43].

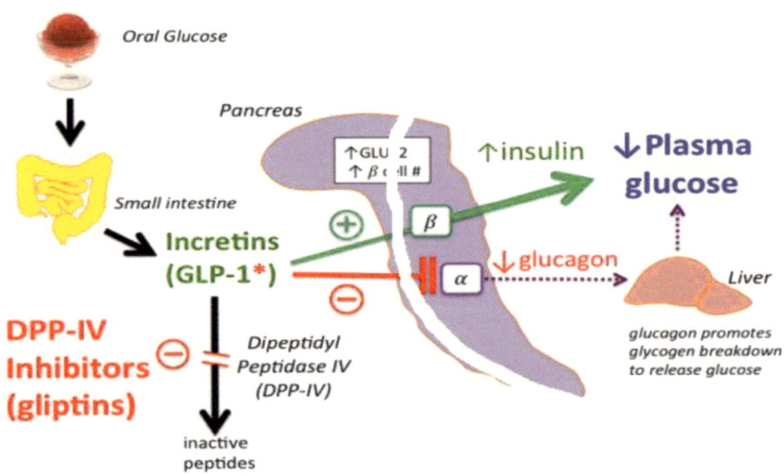

Figure 8.
Metabolic effects of DDP-4 inhibitors.

It has been observed that polymorphism in DPP4 gene is concerned with different cardio-metabolic disorders and transmission of MERS-CoV [44]. The polymorphism in DPP4 reduces the interactions between S proteins of MERS-CoV with cellular membranes. The difference in the incidence between Arabian and African MERS-CoV is mainly related to the genotype polymorphism of DPP4 [45].

Since there is a higher genomic similarity between SARS-CoV-2 and SARS-CoV, the implication of DPP4 as a receptor or a pathogenetic pathway in COVID-19-induced ARDS is reasonable. In COVID-19 infection, ARDS is developed due to massive cytokine release (cytokine storm), due to uncontrolled systemic inflammatory response, and due to the release of pro-inflammatory cytokine, including INF-α-, IL-6, IL-12, IL-33, etc., which cause multiple organ failure [46].

It has been reported that DPP4 is highly expressed on alveolar cells (type I and type II), alveolar macrophage, pleural mesothelium, and vascular endothelium. Besides, pulmonary vascular endothelial cells (PVECs) are the main source of pro-inflammatory cytokines in ARDS. DPP4 is upregulated in pneumonia, asthma, and ARDS; therefore, DPP4 inhibitors may reduce the inflammatory reaction and cytokine release in acute lung injury and ARDS through immune-modulation effect [47].

Different studies illustrated that DPP4 inhibitors inhibit the release of IL-6 and TNF-α in ARDS without effect on the blood glucose when used in a small dose [48]. Therefore, DPP4 inhibitors might be a therapeutic option in the management of COVID-19-induced ARDS.

5. The interactions between RAS and DPP4 in COVID-19

DPP4 is involved in the activation of T-cell, so the activated T-cell can synthesize and secret AngII by local RAS. So, AngII-dependent T-cell activation is inhibited by DPP4i, leading to a significant reduction of endothelial inflammation and vasoconstriction [49]. Similarly, Bengsch et al. [50] disclosed that DPP4 receptors are highly expressed on Th17 which induced AngII release during ARDS. High circulating AngII level leads to the activation of inflammatory reactions via activation of the release of IL-18, 1 L-17, IL-6, and TNF-α. These mediators are essential for the induction of epithelial and endothelial cell injury during ARDS. DPP4i plays an integral role in the attenuation of IL-6 and TNF-α and associated inflammation and endothelial damage [51].

Therefore, DPP4i interferes with RAS at different levels; as teneligliptin attenuates AngII action, liraglutide downregulates AT1 and upregulates AT2 receptors during intravenous AngII infusion. Besides, exenatide inhibits the secretion of renin, AngII, and angiotensinogen [52].

What is more, DPP4i has a nephroprotective effect through the regulation of sodium and water reabsorption. Both DPP4i and GLP-1 increase sodium excretion through direct effect on proximal renal tubules or indirect effect via the activation of brain natriuretic peptide (BNP) (**Figure 9**) [53].

Consequently, DPP4i exerts protective effects on the lung and kidney which are the main tropism of COVID-19 through direct modulation of inflammatory reactions or indirectly through the attenuation of AngII. The interaction between DPP4i and RAS inhibitors seems to augment the expression of AT2 receptor which is under extensive researches to find the pathophysiological pathway of COVID-19 infection. Indeed, this interaction sheds light for possible attenuation of COVID-19-induced ARDS and AKI mainly in critically ill patients with systemic hypertension.

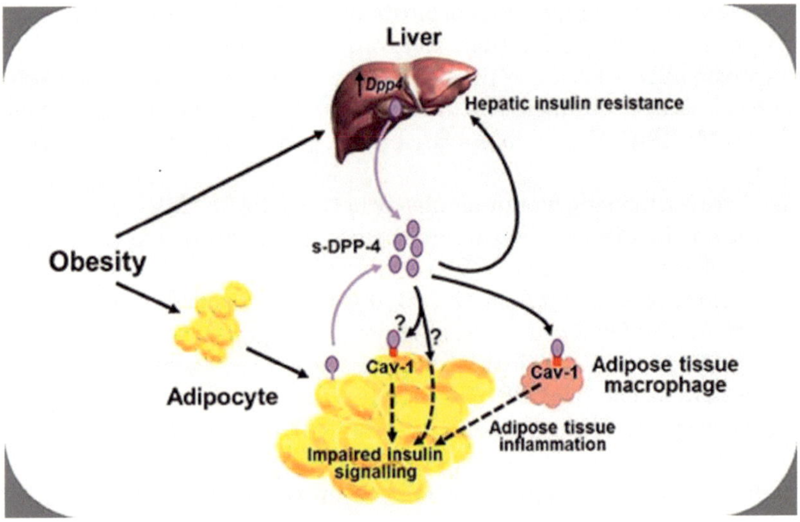

Figure 9.
Metabolic effects of DDP-4 inhibitors regarding obesity and insulin resistance.

6. Conclusion

ACE2 is regarded as a portal entry-point for SARS-CoV-2; however, human cell lines with higher expression of ACE2 are not infected by this virus. The overexpression of AT2 receptors by ACE2 and Ang1–7 is regarded as a novel pathway in COVID-19-induced pneumonia and ARDS. Females have lower levels of ACE2 than males, which give a clue of male vulnerability to COVID-19 infection as compared with females. Even so, ACE2 is regarded as a protective pathway which reduces COVID-19-induced acute inflammatory reactions. RAS inhibitors and DPP4i increase the expression of pulmonary ACE2, so the implication of RAS inhibitors and DPP4i as augmenters of COVID-19 is not practical. Therefore, addition of, DPP4i to RAS inhibitors in hypertensive patients with COVID-19 may reduce the risk of ARDS and AKI.

Conflict of interest

Nil.

Funding and sponsorship

Nil.

Author details

Hayder M. Al-Kuraishy[1*], Marwa S. Al-Niemi[1], Nawar R. Hussain[2], Ali I. Al-Gareeb[1], Nasser A. Al-Harchan[2] and Azhar H. Al-Kurashi[1]

1 Department of Clinical Pharmacology, Medicine and Therapeutic, Medical Faculty, College of Medicine, Al-Mustansiriya University, Baghdad, Iraq

2 Department of Clinical Pharmacology, Medicine and Therapeutic, Medical Faculty, College of Medicine, Al-Farahedi University, Baghdad, Iraq

*Address all correspondence to: hayderm36@yahoo.com

References

[1] Al-kuraishy HM, Al-Maiahy TJ, Al-Gareeb AI, Musa RA, Ali ZH. COVID-19 pneumonia in an Iraqi pregnant woman with preterm delivery. Asian Pacific Journal of Reproduction. 2020;**9**(3):156-167

[2] Peiris JS, Lai ST, Poon LL, Guan Y, Yam LY, Lim W, et al. Coronavirus as a possible cause of severe acute respiratory syndrome. The Lancet. 2003;**361**(9366):1319-1325

[3] Zhong NS, Zheng BJ, Li YM, Poon LL, Xie ZH, Chan KH, et al. Epidemiology and cause of severe acute respiratory syndrome (SARS) in Guangdong, People's Republic of China, in February, 2003. The Lancet. 2003;**362**(9393):1353-1358

[4] Holmes KV. SARS-associated coronavirus. New England Journal of Medicine. 2003;**348**(20):1948-1951

[5] Li Q, Guan X, Wu P, Wang X, Zhou L, Tong Y, et al. Early transmission dynamics in Wuhan, China, of novel coronavirus–infected pneumonia. New England Journal of Medicine. 2020;**29**:133-139

[6] Al-Kuraishy HM, Al-Gareeb AI. From SARS-CoV to nCoV-2019: Ruction and argument. Archives of Clinical Infectious Diseases. 2020;**15**(2):e102624

[7] Brian DA, Baric RS. Coronavirus genome structure and replication. In: Coronavirus Replication and Reverse Genetics. Berlin, Heidelberg: Springer; 2005. pp. 1-30

[8] Sheahan TP, Sims AC, Graham RL, Menachery VD, Gralinski LE, Case JB, et al. Broad-spectrum antiviral GS-5734 inhibits both epidemic and zoonotic coronaviruses. Science Translational Medicine. 2017;**9**(396):1237-1244

[9] Pruijssers AJ, Denison MR. Nucleoside analogues for the treatment of coronavirus infections. Current Opinion in Virology. 2019;**35**:57-62

[10] Kuba K, Imai Y, Rao S, Gao H, Guo F, Guan B, et al. A crucial role of angiotensin converting enzyme 2 (ACE2) in SARS coronavirus–induced lung injury. Nature Medicine. 2005;**11**(8):875-879

[11] Li W, Moore MJ, Vasilieva N, Sui J, Wong SK, Berne MA, et al. Angiotensin-converting enzyme 2 is a functional receptor for the SARS coronavirus. Nature. 2003;**426**(6965):450-454

[12] Eckerle I, Corman VM, Müller MA, Lenk M, Ulrich RG, Drosten C. Replicative capacity of MERS coronavirus in livestock cell lines. Emerging Infectious Diseases. 2014;**20**(2):276

[13] Pinter M, Jain RK. Targeting the renin-angiotensin system to improve cancer treatment: Implications for immunotherapy. Science Translational Medicine. 2017;**9**(410):eaan5616

[14] Santos RA, Sampaio WO, Alzamora AC, Motta-Santos D, Alenina N, Bader M, et al. The ACE2/angiotensin-(1-7)/MAS axis of the renin-angiotensin system: Focus on angiotensin-(1-7). Physiological Reviews. 2018;**98**(1):505-553

[15] Wan Y, Shang J, Graham R, Baric RS, Li F. Receptor recognition by the novel coronavirus from Wuhan: An analysis based on decade-long structural studies of SARS coronavirus. Journal of Virology. 2020;**94**(7):71-79

[16] Fang L, Karakiulakis G, Roth M. Are patients with hypertension and diabetes mellitus at increased risk for COVID-19

infection? The Lancet Respiratory Medicine. 2020;**11**:39-51

[17] Cai H. Sex difference and smoking predisposition in patients with COVID-19. The Lancet Respiratory Medicine. 2020;**11**:22-29

[18] Zhou P, Yang XL, Wang XG, Hu B, Zhang L, Zhang W, et al. Discovery of a novel coronavirus associated with the recent pneumonia outbreak in humans and its potential bat origin. BioRxiv. 2020. Preprint. DOI: 10.20944/preprints202003.0295.v1

[19] Li W, Sui J, Huang IC, Kuhn JH, Radoshitzky SR, Marasco WA, et al. The S proteins of human coronavirus NL63 and severe acute respiratory syndrome coronavirus bind overlapping regions of ACE2. Virology. 2007;**367**(2):367-374

[20] Tian X, Li C, Huang A, Xia S, Lu S, Shi Z, et al. Potent binding of 2019 novel coronavirus spike protein by a SARS coronavirus-specific human monoclonal antibody. Emerging Microbes & Infections. 2020;**9**(1):382-385

[21] Qi Y, Shenoy V, Wong F, Li H, Afzal A, Mocco J, et al. Lentivirus-mediated overexpression of angiotensin-(1-7) attenuated ischaemia-induced cardiac pathophysiology. Experimental Physiology. 2011;**96**(9):863-874

[22] Xiaopeng L, Zhuang J, Uhal BD. Local activation of the pulmonary extravascular angiotensin system induces epithelial apoptosis and lung fibrosis. Journal of Lung Cancer, Pulmonary & Respiratory Research. 2018;**5**(6):192-200

[23] Annoni F, Orbegozo D, Rahmania L, Irazabal M, Mendoza M, De Backer D, et al. Angiotensin-converting enzymes in acute respiratory distress syndrome. Intensive Care Medicine. 2019;**45**(8):1159-1160

[24] Palau V, Riera M, Roca-Ho H, Benito D, Gimeno J, Pascual J, et al. MP498 ANGIOTENSIN CONVERTING ENZYME 2 DELETION INCREASES ACE EXPRESSION AND OXIDATIVE STRESS IN PANCREAS FROM NOD MICE. Nephrology Dialysis Transplantation. 2017;**32**(3):iii612

[25] Li G, Zhang H, Zhao L, Zhang Y, Yan D, Liu Y. Angiotensin-converting enzyme 2 activation ameliorates pulmonary endothelial dysfunction in rats with pulmonary arterial hypertension through mediating phosphorylation of endothelial nitric oxide synthase. Journal of the American Society of Hypertension. 2017;**11**(12):842-852

[26] Kuba K, Imai Y, Ohto-Nakanishi T, Penninger JM. Trilogy of ACE2: A peptidase in the renin-angiotensin system, a SARS receptor, and a partner for amino acid transporters. Pharmacology & Therapeutics. 2010;**128**(1):119-128

[27] Wang PH. Increasing host cellular receptor-angiotensin-converting enzyme 2 (ACE2) expression by coronavirus may facilitate 2019-nCoV infection. bioRxiv. 2020

[28] Al-kuraishy HM, Al-Gareeb AI, Al-nami MS. Irbesartan attenuates gentamicin-induced nephrotoxicity in rats through modulation of oxidative stress and endogenous antioxidant capacity. International Journal of Preventive Medicine. 2020;**11**

[29] Zhang H, Penninger JM, Li Y, Zhong N, Slutsky AS. Angiotensin-converting enzyme 2 (ACE2) as a SARS-CoV-2 receptor: Molecular mechanisms and potential therapeutic target. Intensive Care Medicine. 2020;**3**:1-5

[30] Meng Y, Pan M, Zheng B, Chen Y, Li W, Yang Q, et al. Autophagy attenuates angiotensin II-induced pulmonary fibrosis by inhibiting redox

imbalance-mediated NOD-like receptor family pyrin domain containing 3 inflammasome activation. Antioxidants & Redox Signaling. 2019;**30**(4):520-541

[31] Qiu Y, Zhao YB, Wang Q, Li JY, Zhou ZJ, Liao CH, et al. Predicting the angiotensin converting enzyme 2 (ACE2) utilizing capability as the receptor of SARS-CoV-2. Microbes and Infection. 2020;**19**:45-52

[32] Gu J, Korteweg C. Pathology and pathogenesis of severe acute respiratory syndrome. The American Journal of Pathology. 2007;**170**(4):1136-1147

[33] Gurwitz D. Angiotensin receptor blockers as tentative SARS-CoV-2 therapeutics. Drug Development Research. 2020;**4**:70-79

[34] Procko E. The sequence of human ACE2 is suboptimal for binding the S spike protein of SARS coronavirus 2. bioRxiv. 2020. Preprinted. DOI: 10.1101/2020.03.16.994236

[35] Corley MJ, Ndhlovu LC. DNA methylation analysis of the COVID-19 host cell receptor, angiotensin I converting enzyme 2 gene (ACE2) in the respiratory system reveal age and gender differences. Gender. Preprint. DOI: 10.20944/preprints202003.0295.v1

[36] Shenoy V, Grobe JL, Qi Y, Ferreira AJ, Fraga-Silva RA, Collamat G, et al. 17β-Estradiol modulates local cardiac renin-angiotensin system to prevent cardiac remodeling in the DOCA-salt model of hypertension in rats. Peptides. 2009;**30**(12):2309-2315

[37] Gallagher PE, Ferrario CM, Tallant EA. Regulation of ACE2 in cardiac myocytes and fibroblasts. American Journal of Physiology-Heart and Circulatory Physiology. 2008;**295**(6):H2373-H2379

[38] Haga S, Nagata N, Okamura T, Yamamoto N, Sata T, Yamamoto N, et al. TACE antagonists blocking ACE2 shedding caused by the spike protein of SARS-CoV are candidate antiviral compounds. Antiviral Research. 2010;**85**(3):551-555

[39] Berger JP, SinhaRoy R, Pocai A, Kelly TM, Scapin G, Gao YD, et al. A comparative study of the binding properties, dipeptidyl peptidase-4 (DPP-4) inhibitory activity and glucose-lowering efficacy of the DPP-4 inhibitors alogliptin, linagliptin, saxagliptin, sitagliptin and vildagliptin in mice. Endocrinology, Diabetes & Metabolism. 2018;**1**(1):e00002

[40] Pantaleão SQ, Philot EA, Resende-Lara D, Túlio P, Lima AN, Perahia D, et al. Structural dynamics of DPP-4 and its influence on the projection of bioactive ligands. Molecules. 2018;**23**(2):490

[41] White WB, Wilson CA, Bakris GL, Bergenstal RM, Cannon CP, Cushman WC, et al. Angiotensin-converting enzyme inhibitor use and major cardiovascular outcomes in type 2 diabetes mellitus treated with the dipeptidyl peptidase 4 inhibitor alogliptin. Hypertension. 2016;**68**(3):606-613

[42] Widagdo W, Raj VS, Schipper D, Kolijn K, van Leenders GJ, Bosch BJ, et al. Differential expression of the Middle East respiratory syndrome coronavirus receptor in the upper respiratory tracts of humans and dromedary camels. Journal of Virology. 2016;**90**(9):4838-4842

[43] Raj VS, Mou H, Smits SL, Dekkers DH, Müller MA, Dijkman R, et al. Dipeptidyl peptidase 4 is a functional receptor for the emerging human coronavirus-EMC. Nature. 2013;**495**(7440):251-254

[44] Kleine-Weber H, Schroeder S, Krüger N, Prokscha A, Naim HY, Müller MA, et al. Polymorphisms

in dipeptidyl peptidase 4 reduce host cell entry of Middle East respiratory syndrome coronavirus. Emerging Microbes & Infections. 2020;**9**(1):155-168

[45] Shirato K, Melaku SK, Kawachi K, Nao N, Iwata-Yoshikawa N, Kawase M, et al. Middle East respiratory syndrome coronavirus in dromedaries in Ethiopia is antigenically different from the Middle East isolate EMC. Frontiers in Microbiology. 2019;**10**:1326

[46] Huang C, Wang Y, Li X, Ren L, Zhao J, Hu Y, et al. Clinical features of patients infected with 2019 novel coronavirus in Wuhan, China. The lancet. 15 February 2020;**395**(10223):497-506

[47] Vliegen G, Raju TK, Adriaensen D, Lambeir AM, De Meester I. The expression of proline-specific enzymes in the human lung. Annals of Translational Medicine. 2017;**5**(6):41-53

[48] Li H, Zhang J, Lin L, Xu L. Vascular protection of DPP-4 inhibitors in retinal endothelial cells in in vitro culture. International Immunopharmacology. 2019;**66**:162-168

[49] Shah Z, Kampfrath T, Deiuliis JA, Zhong J, Pineda C, Ying Z, et al. Long-term dipeptidyl-peptidase 4 inhibition reduces atherosclerosis and inflammation via effects on monocyte recruitment and chemotaxis. Circulation. 2011;**124**(21):2338-2349

[50] Bengsch B, Seigel B, Flecken T, Wolanski J, Blum HE, Thimme R. Human Th17 cells express high levels of enzymatically active dipeptidylpeptidase IV (CD26). The Journal of Immunology. 2012;**188**(11):5438-5447

[51] Esposito G, Cappetta D, Russo R, Rivellino A, Ciuffreda LP, Roviezzo F, et al. Sitagliptin reduces inflammation, fibrosis and preserves diastolic function in a rat model of heart failure with preserved ejection fraction. British Journal of Pharmacology. 2017;**174**(22):4070-4086

[52] Chaudhuri A, Ghanim H, Makdissi A, Green K, Abuaysheh S, Batra M, et al. Exenatide induces an increase in vasodilatory and a decrease in vasoconstrictive mediators. Diabetes, Obesity and Metabolism. 2017;**19**(5):729-733

[53] Boerrigter G, Costello-Boerrigter LC, Harty GJ, Lapp H, Burnett JC Jr. Des-serine-proline brain natriuretic peptide 3-32 in cardiorenal regulation. American Journal of Physiology: Regulatory, Integrative and Comparative Physiology. 2007;**292**(2):R897-R901